Erwin Karer

Subspace Correction Methods for Linear Elasticity

Erwin Karer

Subspace Correction Methods for Linear Elasticity

Algebraic Multigrid & Subspace Corrections for almost incompressible Materials

Südwestdeutscher Verlag für Hochschulschriften

Impressum / Imprint
Bibliografische Information der Deutschen Nationalbibliothek: Die Deutsche Nationalbibliothek verzeichnet diese Publikation in der Deutschen Nationalbibliografie; detaillierte bibliografische Daten sind im Internet über http://dnb.d-nb.de abrufbar.

Bibliographic information published by the Deutsche Nationalbibliothek: The Deutsche Nationalbibliothek lists this publication in the Deutsche Nationalbibliografie; detailed bibliographic data are available in the Internet at http://dnb.d-nb.de.

Coverbild / Cover image: www.ingimage.com

Verlag / Publisher:
Südwestdeutscher Verlag für Hochschulschriften
ist ein Imprint der / is a trademark of
AV Akademikerverlag GmbH & Co. KG
Heinrich-Böcking-Str. 6-8, 66121 Saarbrücken, Deutschland / Germany
Email: info@svh-verlag.de

Herstellung: siehe letzte Seite /
Printed at: see last page
ISBN: 978-3-8381-3103-0

Zugl. / Approved by: Linz, Johannes Kepler Universität Linz, Dissertation, 2011

Abstract

The framework of subspace correction methods describes approaches to solve finite element discretizations of elliptic partial differential equations. Examples of efficient solution techniques are multigrid methods, domain decomposition methods, and auxiliary space methods.

As a first result, we derive the error norm of the method of successive subspace corrections in case of two subspaces using strengthened Cauchy-Bunyakovsky-Schwarz inequalities to estimate energy minimizing restrictions of the operator on subspaces.

Next, we focus on the system of elliptic partial differential equations modeling the stresses and displacements in linear elastic materials in primal variables. There are two basic approaches to set up a variational framework for such models. On one hand there is a mixed formulation resulting in indefinite linear systems of algebraic equations for the discrete solution. On the other hand, there is a formulation in primal variables, which gives rise to a symmetric positive (semi)-definite discrete problem.

First, we consider the standard discretization of the linear elasticity equations by means of continuous piecewise linear finite elements. This discretization suffers from volume locking as the material becomes nearly incompressible. We first consider the case in which such low order conforming methods provide sufficiently accurate approximation to the displacement field. It is well known that the classical algebraic multigrid (AMG) methods do not perform well on this problem without modifications.

We study one competitive AMG method for solving the symmetric positive definite system resulting from the discretization of the elasticity problem. In this method, the coarsening is based on so-called edge matrices, which allows to generalize the concept of strong and weak connections, as used in classical AMG, to "algebraic vertices" that accumulate the nodal degrees of freedom in case of vector-field problems. The major contribution is devising a measure for the nodal dependence which guides the generation of the edge matrices, which are the basic building blocks of this method. A natural measure is the abstract angle between the two subspaces spanned by the basis functions corresponding to the vertices forming an edge in our finite element partition. Another original contribution of this work is a two-level convergence analysis of the method. The presented numerical results cover also problems with jumps in the Young's modulus of elasticity and orthotropic materials, like wood or cancellous bone.

In a second part, we investigate the equations of elasticity in primal variables for nearly incompressible materials, like rubber. For such materials, i.e., when the first Lamé parameter tends to infinity this problem becomes ill-posed and the resulting discrete problem is nearly singular.

Due to the locking of approximations using conforming, low order polynomial spaces, to obtain any meaningful approximation to the displacement field, one has to use finite element spaces of at least order four (or even higher). Alternatively, and this is what we are studying here, one can consider stable nonconforming finite element discretizations based on reduced integration. One main question which then arises, and which we address here is how to construct a robust (uniform in the problem parameters, such as Lamé's first parameter) iterative solution method for the resulting system of linear algebraic equations. We introduce a specific space decomposition into two overlapping subspaces that serves as a basis for devising a uniformly convergent subspace correction algorithm. The first subspace consists of weakly divergence-free functions. The second subspace is the complementary space which we augment with a suitably chosen overlap by adding certain weakly divergence free components. We solve the two subproblems exactly. This subspace correction method gives rise to a preconditioner which is a convex combination of a multiplicative preconditioner (based on the subspace splitting we mentioned above) plus a solution of a system equivalent to vector Laplace equation (for which efficient methods exist). We present a pool of numerical tests confirming the uniform convergence.

Acknowledgements

First of all, I express great appreciation to my advisor Johannes Kraus, for introducing me to the field of algebraic multigrid and multilevel methods, for supervising this thesis, for useful hints and interesting discussions in an always friendly environment. Moreover, I owe a lot of gratitude to Ludmil Zikatanov for co-refereeing the thesis and for his ability to explain difficult matters in a simple way to me.

Special thanks goes to my colleagues Astrid and Clemens Pechstein, Michael Kolmbauer, Markus Kollmann, Sven Beuchler, Jörg Willems, Huidong Yang, Ivan Georgiev, Martin Huber and Martin Purrucker for the good and stimulating working atmosphere. They always helped me when I have had hard problems to solve.

Next, I acknowledge the great scientific environment provided by H. W. Engl, the director of RICAM, Austrian Academy of Sciences, and by U. Langer, the leader of the group on direct fields methods. Additionally, I would like to say thank you to the administrative staff of RICAM for providing me with all the help I needed through the years.

My deepest thanks goes to Tanja for all her love, emotional support and patience. Further, I want to express my sincere appreciation to my family for all their help and encouragements when it was necessary.

Last, but not least, I acknowledge the financial support given by the Austrian Academy of Sciences and by the Austrian Science Fund (FWF) through the project P19170-N18.

Contents

Chapter 1

Introduction

1.1 State of the art

Nowadays, the simulation of mechanical problems arises in many different fields of research and engineering. Examples are strength and stiffness calculations of certain (parts of) machines or devices, or crash simulations of cars. The system of mechanical equations might also couple with other systems of physical equations. In medicine for instance, fluid structure interaction is used to simulate the blood flow through blood vessels. There the forces acting on the vessels stem from the flow of blood through the artery or vein.

Physical models are typically formulated in terms of *Partial Differential Equations* (PDEs). The equations of elasticity describe the deformations and stresses of an elastic body under applied volume and surface forces. The balance of momentum implies the steady state relation between the stress field and the forces. Further, the stress and the strain are coupled via material laws. The simplest one is *Hooke's law*, which states a linear dependence of the stress on the strain. This law is valid for elastic materials and small deformations, that is, as long as the load does not exceed the material's "elastic limit". The arising equations are nonlinear. Nevertheless, the assumption of small deformations allows for a linearization of the equations and leads to the equations of linear elasticity. Note that even for larger deformations or non-linear material laws an in-depth understanding of the linearized system is important.

In physics, PDEs often stem from a variational formulation, which arises from minimizing the energy of a given system. Under suitable assumptions the equivalence of both descriptions can be shown. If the system of linear elasticity equations is expressed solely in terms of the displacements, its variational formulation is known as *classical* or *pure displacement formulation.*

Mathematically, the variational problem is set up in the Sobolev space H^1. If the material under consideration becomes almost incompressible, the compliance tensor, which links stress and strain, deteriorates, i.e., it becomes ill-conditioned with increasing the incompressibility. In the incompressible limit it is singular. In this case its inverse is not well-defined. Since the inverse of the compliance tensor directly enters the pure displacement formulation we observe instabilities in the incompressible limit. This effect is called *volume locking* or sometimes simply *locking*. A remedy to locking is to pose the elasticity problem variationally via so-called *mixed formulations*. There, either the stress or the pressure enters the model as an additional variable. These formulations can be shown to be stable with respect to incompressibility. In the incompressible limit one still needs certain compatibility conditions for mixed formulations due to the singularity of the compliance tensor.

In general, it is not possible to solve the considered equations in two or three dimensions and on general domains analytically. Therefore, a numerical approximation of the solution is of interest. The most commonly used method is the *finite element method* (FEM). Thereby, the domain is partitioned into small, usually polygonal, subdomains. Those subdomains are called elements of the *mesh*. The most common elements are triangles or quadrilaterals in two space dimensions (2D) and tetrahedra or hexahedra in 3D. On each element we approximate the solution by polynomials, leading to a global piecewise polynomial function. In the case of conforming finite element spaces, the global functions have to be continuous. In this case existence and uniqueness of the (numerical) solution of the discretized (finite dimensional) problem directly follow from the existence and uniqueness of the solution of the (infinite-dimensional) continuous problem.

If the pure displacement formulation is discretized by means of conforming piecewise polynomials we observe a lack of accuracy with respect to the incompressibility when polynomials of order less than four are used. However, if the finite element space is of order 4 or higher optimal order estimates can be shown robustly with respect to incompressibility ([SV85]). On the other hand, for lower order conforming spaces, employing the technique of *reduced integration* (cf. [ESGB82, HLB79]) yields optimal order error estimates (see [Fal91]) for isotropic materials in the case of plane stress (2D). Thereby, the integral involving the divergence operator is approximated by a reduced order quadrature rule. In [Sch99b, Sch99a] it is shown that this formulation using first-order elements is equivalent to a mixed formulation for displacements and pressure in the spaces of piecewise $\mathcal{P}_2$- and $\mathcal{P}_0$-functions, respectively. This pair of spaces is known to be stable for Stokes equations. Indeed, stable pairs of spaces of mixed FEM for Stokes equations are usually locking free (see [GR86, BF91]).

An adaption of FEM is the so-called *discontinuous Galerkin* (DG) finite element method.

Thereby, the finite element spaces are nonconforming, i.e., they consist of piecewise polynomial functions that do not have to be continuous. The discontinuities of the solution across element interfaces, in short jumps, are penalized by adding certain penalty terms to the arising bilinear form which then do imply coercivity of the bilinear form. Note that due to the discontinuity of the space the DG approximations involve a much higher number of degrees of freedom as the related conforming approximations. There do exist stable DG discretizations of the system of linear elasticity equations with respect to incompressibility (cf. [HL03]).

The discrete variational problem allows then to set up a system matrix A and right-hand side vector. In our case they represent the mechanical properties of the deformed body and the acting forces. Using specially constructed direct solution methods it requires at best $\mathcal{O}(n^2)$ (see [GL81]) floating point operations to solve the linear algebraic system of equations of size n for problems in three dimensions, which is not desirable if n is extremely large. Therefore, iterative solution methods are used. Thereby, we say that a method is of optimal order if it reaches the desired accuracy with $\mathcal{O}(n)$ floating point operations. Quite often methods are quasi-optimal which means that $\mathcal{O}(n \log(n))$ operations are needed. Solvers or preconditioners are said to be *robust* if problem as well as discretization parameters do not influence their convergence behavior qualitatively. In elasticity for instance, the problem parameters are the Young's modulus of elasticity and the Poisson ratio. Typically, discretization parameters are the size of the triangulation and the used polynomial order.

The convergence rate of classical stationary iterative solution methods like Richardson, Jacobi or Gauss-Seidel iteration deteriorates with decreasing mesh size. The method of *conjugate gradients* (CG), see [HS52, Axe94], is a method of choice for solving large sparse symmetric positive definite systems of linear algebraic equations. Its convergence rate depends on the square root of the condition number of the considered matrix. Since the condition number for the considered system of equations is of the order h^{-2}, where h denotes the mesh size of the triangulation, it is not robust. However, if we are able to find a spectrally equivalent preconditioner for the (stiffness) matrix, such that the condition number of the preconditioned system stays bounded when h tends to zero, we can solve the linear system up to any prescribed (but fixed) accuracy with a bounded number of preconditioned CG iterations. Finally, the total procedure is of optimal order if the setup and the application of the preconditioner are of the complexity $\mathcal{O}(n)$.

The first developed solution procedures of optimal order are *multigrid* (MG) methods (see for instance [HT82, Hac85, TOS01]). As the name indicates multigrid is based on a hierarchy of grids. They rely on the two complementary concepts of *smoothing* and *coarse grid correction.* First, a fast relaxation procedure is applied on the fine grid in order to eliminate the high

frequency components of the error. Then, the remaining smooth error can be accurately represented on the next coarser grid. There, the coarse grid correction step removes the remaining error components. This can be either done by solving the problem on the coarse level exactly, or by applying the MG procedure recursively, i.e., error smoothing on the current grid and coarse grid correction on the next coarser grid. The basis for the MG method to be of optimal order is an excellent concurrence between smoothing and coarse grid correction. That is, error components not being reduced by the relaxation have to be efficiently reduced by the coarse grid correction and vice versa. This is reflected in the convergence analysis of Hackbusch [Hac85] which is based on the *smoothing* and *approximation property*. Another crucial point in the design of MG methods is a proper choice of the transfer operators. Thereby, error components in the (near-)kernel of the system operator have to be represented well on the coarser grids, that is, the transfer operator has to be *kernel preserving*.
Multigrid methods have been successfully applied to the pure displacement formulation of linear elasticity (see [KM87] for instance). Additionally, in [Sch99b, Sch99a] a robust MG method is constructed for the pure displacement problem based on reduced integration.

The *algebraic multigrid* (AMG) method has been developed aiming at the superior properties of MG methods, i.e, the optimality, but being not dependent on the existence of a hierarchy of meshes. Usually, a smoother is chosen first. Then, a hierarchy of algebraic meshes is set up based on algebraic considerations using the matrix A. In standard AMG (e.g. [BMR82, RS87]) the *coarse grid selection* uses the *strength of connectivity* measuring the rate of dependence of single DOF on each other. This measure is additionally used to determine the interpolation operator. As for standard MG methods the smoother and the coarse grid correction have to to complement each other. That is, *algebraically smooth error*, i.e., error components that cannot be removed efficiently by the smoother have to be reduced by coarse-grid correction. The classical AMG [RS87] does not perform well for elasticity problems, because its construction of the interpolation is based on the assumption of only preserving constants.

Another, purely algebraic AMG variant is *smoothed aggregation* [VBM01, VMB96], which uses aggregates of fine DOF as coarse of. The piecewise constant interpolation is improved by a proper smoothing. The *adaptive AMG* [BFM$^+$06] and *adaptive smoothed aggregation AMG* [BFM$^+$04] detect the smooth error, with respect to a certain smoother S, during the process and improve their own components. In [BFM$^+$04] the kernel of problem can be a-priori be provided as an input which improves the convergence behavior. Additionally, satisfactory results for linear elasticity problems have been reported in [BFM$^+$04].

A different type of AMG methods uses certain additional information on the FE discretization in the set-up phase. In the so-called AMGe methods this information is extarcted from the element stiffness matrices[BCF$^+$01, HV01, JV01, CFH$^+$03]. In the recent works [Kra08, KK10] additionally the vertex coordinates of the triangulation are used to set up the coarse grids and the interpolation operators for solving the pure displacement problem of linear elasticity. Finally, the highly parallel and purely algebraic AMG variant *BoomerAMG* [HMY00] together with an improved parallel coarsening algorithm [DSMYH06] is one example of a highly efficient linear solver for different types of elliptic problems.

Another type of methods that are applicable also to discretizations of linear elasticity problems is preconditioning by *algebraic multilevel iteration* (AMLI) [KM09, Vas08], or the original papers [AV89, AV90]. Thereby, a hierarchical basis of the discrete space can be exploited (similar to the hierarchical basis multigrid ([Yse86, BDY87]). In AMLI, the multilevel procedure is stabilized by employing certain polynomials, e.g., Chebyshev polynomials.

Finally, there is the class of *domain decomposition* (DD) methods for solving discretizations of PDEs. The general framework of DD methods is a decomposition of the computational domain into a set of much smaller subdomains. On each subdomain, the problem is easier solvable due to the reduced size. If the subdomains are not disjoint, the DD methods are referred to as *Schwarz overlapping DD methods* [TW05]. If the decomposition is a partition of the domain the DD method is called *substructuring* DD method, for which the most prominent procedures are Neumann-Neumann, finite element tearing and interconnecting (FETI), FETI-DP and balancing domain decomposition by constraints (BDDC) methods. For more details on DD methods and their application to linear elasticity we refer to [TW05, Pec08, Pec12] and the references given therein.

All the previous methods belong to the class of *subspace correction methods*. The term subspace correction was introduced in [Xu92]. In general, subspace correction methods are dealing with uniquely solvable variational problems on a Hilbert space V. This space is divided into subspaces on which the residual equation is solved in parallel (*method of parallel subspace corrections* (MPSC)) or sequentially (*method of successive subspace corrections* (MSSC)) in each step. Those methods can be viewed as (overlapping) block-Jacobi or block Gauss-Seidel methods. Their theoretical framework aligns with Schwarz methods (see [GO95, TW05]). In [XZ02] an identity for the norm of the error propagation of MSSC is shown. The theory has been further investigated in [Zik08, CXZ08].

1.2 On this work

The aim of this book is to provide efficient preconditioners for the equations of linear elasticity. Thereby, efficiency means that we target on robustness with respect to problem and discretization parameters, and further on optimality with respect to the problem size. We focus on the pure displacement variational formulation of linear elasticity. In this case the problem parameters are the *Lamé* parameters λ and μ, whereas special attention has to be drawn to λ. In the case of almost incompressible materials λ becomes very large, which leads to an ill-conditioned system of algebraic equations. Since classical (conforming) finite element approximations suffer from volume locking when used as a discretization tool for approximating the displacement field of nearly incompressible elastic bodies we employ them only in the compressible regime.

In the first part of this work we study a specific AMG method called *algebraic multigrid based on computational molecules* (AMGm) which has been developed in [KS06] and enhanced in [Kra08] for problems in linear elasticity discretized by piecewise linear elements. This method uses so-called "edge matrices" in order to represent the coupling between the DOF of two vertices. These matrices are used to set up "computational molecules" in order to determine the strength of connectivity. Additionally, computational molecules, assembled from edge matrices, are used to define interpolation.

We enhance and improve this procedure here as follows. First, we generalize the concept and introduce "algebraic vertices" and "algebraic edges". For each algebraic edge we set up an edge matrix representing the dependence of the two algebraic vertices it connects. Then we develop a refined measure for the strength of connectivity. For a given edge, the new measure is defined as the energy cosine of the abstract angle between the two spaces spanned by the respective basis functions associated with the two vertices that are connected via the edge. While the interpolation routine remains the same, we use the knowledge of the structure of the edge matrices to obtain better approximations via those matrices. We also improve the computation of the coarse edge matrices which finally leads to a much better convergence behavior of the improved method compared to the original AMGm procedure. In the numerical experiments we apply AMGm to different types of elasticity problems showing its robustness with respect to the specific parameters. Moreover, we compare the method to its predecessor and to the state-of-the-art BoomerAMG [HMY00, DSMYH06]. In the comparison we see that the method outperforms its predecessor as well as it outperforms BoomerAMG in terms of convergence. Nevertheless, this has to be accounted for by a higher complexity (and more expensive setup phase). Additionally, we discuss how to parallelize AMGm and moreover, we provide theoretical insight to the method by investigating the two-level method. The main results have been published in [KK10].

In a second part, we focus on almost incompressible materials. As mentioned above, the classical pure displacement formulation suffers from locking in this case. Therefore, we consider a discretization of the pure displacement problem using reduced integration of the divergence-term (cf. [Fal91, Sch99b]). This formulation is shown to result in optimal order error estimates independently of the material parameters. Again we focus on piecewise linear conforming finite elements. In [Sch99b, Sch99a] the equivalence of this problem to a stable $\mathcal{P}_2 - \mathcal{P}_0$ mixed formulation for the Stokes problem is shown.

We decompose the space into the set of *weakly divergence-free* functions and its complement for which the divergence term does not vanish, corresponding to algebraic gradients across edges. The weakly divergence-free space can be further decomposed into proper subspaces. We present locally supported basis functions for all those spaces. This space decomposition is used to set up an overlapping decomposition of the vector space. The overlap is chosen appropriately to be a subspace of weakly divergence-free functions in order to achieve good convergence properties of the MSSC (see [Xu92, XZ02]). Then, the MSSC gives rise to a stand-alone solver and to a uniform preconditioner. In both cases we use the fact that the problem of linear elasticity in the compressible regime can be handled efficiently, which we have demonstrated in the first part of this work! In the scope of this work the discussion has been for exact solvers of the occurring subproblems. In general, the subproblems can be solved by means of efficient preconditioners. On the first subspace, the divergence vanishes and hence the problem is spectrally equivalent to the vector Laplacian problem, which is well-understood. However, the second subproblem is much more difficult to solve since it involves Lamé's first parameter λ. We exploit the *auxiliary space method* (cf. [HX07]) to devise a block-diagonal preconditioner on this subspace. The blocks correspond to a scaled zero-order term and a second term involving the divergence of Raviart-Thomas functions. The zero-order term is easily invertible and efficient solvers for the Raviart-Thomas part are considered in [HX07, KT11]. Finally, this leads to an efficient preconditioner for the considered example. We report the convergence properties of numerical examples. This method is discussed in the accepted article [KKZ11].

Another contribution is the discussion of the convergence behavior of MSSC in case of two overlapping subspaces. The analysis uses the XZ-identity of [XZ02] stating an exact description for the norm of error propagation operator of a general MSSC. We show that in our considered case, this norm is exactly given by the CBS constant of the system operator after eliminating the overlap of the two subspaces. Further, numerical tests confirm the results.

This book is organized as follows:

In *Chapter* 2 we introduce the basic ingredients like function spaces and inequalities on Sobolev spaces which are needed to prove uniqueness of solution of variational problems. Further, we remark on the finite element method and shortly motivate the equations of linear elasticity for which different types of variational formulations are presented.

In *Chapter* 3 we introduce the framework of subspace correction methods and its basic convergence results. Afterwards, different types of subspaces correction methods like (algebraic) multigrid, the auxiliary space method and domain decomposition are briefly described within the subspace correction framework. Finally, the convergence properties in the case of two overlapping subspaces are investigated.

Chapter 4 deals with the new AMG variant for linear elasticity in the pure displacement formulation. The concept of "algebraic vertices" and "algebraic edges" is introduced. Then, the properties on the edge matrices are discussed in order to yield a spectrally equivalent approximation. The basic building blocks of the procedure like the computation of the strength of connectivity, the coarse grid selection process as well as the set up of interpolation operator are discussed in detail. Further, the condition number of the preconditioned system operator is investigated (in the two-level framework). We shortly address parallelization aspects of AMGm. Numerical results are presented for different types of problems and additionally, the method is compared to its predecessor and to the state-of-the-art BoomerAMG. Finally, the application of AMGm to a discontinuous Galerkin formulation of the linear elasticity problem is studied.

In *Chapter* 5 we discuss MSSC applied to the problem stemming from the pure displacement formulation using reduced integration. After introducing some notation and preliminary results, we present a specific splitting of the space of piecewise linear functions based on local basis functions. We utilize this splitting to set up an overlapping space decomposition which is the basis for an MSSC. Next, the preconditioner, which is naturally defined through the MSSC, is introduced. Afterwards, we address how to solve the arising subproblems efficiently. We suggest to use the auxiliary space method to derive a spectrally equivalent preconditioner of one subproblem. Finally, numerical tests are performed to confirm the practicability and efficiency of the proposed method.

Finally, in *Chapter* 6 we summarize the presented results, draw conclusions and give an outlook on possible future work related to this book.

1.3 Notation

Now, let us introduce the symbols and notations we use in this book. Let $\Omega \subset \mathbb{R}^d$ be an open Lipschitz domain ([Gri85]) of dimension $d = 2, 3$. Bold face symbols denote tensors, such as $\boldsymbol{\varepsilon}$, $\boldsymbol{\sigma}$, or vector-valued quantities like $\boldsymbol{u}$ and $\boldsymbol{v}$.

d	spatial dimension, $d = 2, 3$
$\boldsymbol{x}$	spatial coordinate, $\boldsymbol{x} \in \mathbb{R}^d$
$\boldsymbol{v}$	vector or vector-valued function of size $l \in \mathbb{N}$, i.e., $\boldsymbol{u} = (u_i)_{i=1,\dots,l}$
$\boldsymbol{v}^T$, A^T	the transpose of a vector $\boldsymbol{v}$ or of a matrix A
n	number of degrees of freedom (DOFs)
$\partial\Omega$	boundary of the domain Ω; $\partial\Omega = \overline{\Omega}\backslash\Omega$
Γ_D, Γ_N	Dirichlet and Neumann boundary on $\partial\Omega$; see pages 22 and 26
$L^2(D)$	space of square-integrable functions on the domain D, see (2.1)
$\mathcal{C}_0^\infty(\Omega)$	space of analytical functions with compact support on Ω
$\mathcal{D}'(\Omega)$	space of distributions on Ω, i.e., the dual space of $\mathcal{C}_0^\infty(\Omega)$
$H^1(\Omega)$, $H(\text{div};\Omega)$	Sobolev spaces of square-integrable ∇ or div, see (2.4)
$H_0^1(\Omega)$, H_{0,Γ_D}^1	subspaces of $H^1(\Omega)$; the trace vanishes on $\partial\Omega$ or Γ_D, see (2.7),(2.8)
$H_S(\Omega)$	space of symmetric tensors of size $d \times d$ in $L^2(\Omega)$
$\boldsymbol{V}^{\text{RBM}}$, $\hat{\boldsymbol{H}}^1(\Omega)$	space of rigid motions and its complementary space with respect to $[H^1(\Omega)]^d$, see (2.11), (2.12)
H, V	general Hilbert spaces
$(.\,,\,.)_V$	inner product on V inducing a norm $\lVert v\rVert_V := \sqrt{(v\,,\,v)_V}$; the subscript is skipped if the considered space follows from the context.
$a(.,\,.)$	bounded and coercive bilinear form; inducing the energy inner product $(.\,,\,.)_a$; see (2.25) and (2.26)
A	system operator/matrix corresponding to $a(.,\,.)$; see page 38
H'	dual space of H, i.e., the space of linear functionals of H
$L(.)$	linear functional and right-hand side of problem (2.24)
B^t	adjoint of an operator B with respect to $(.,\,.)$
B^*	adjoint of an operator B with respect to $(.,\,.)_a$
I_V	identity operator on a space V; if V is clear from context we simply write I, omitting the subscript
$\lVert\cdot\rVert_B$	norm induced by a symmetric, coercive and bounded linear B

$\boldsymbol{n}$	unit normal vector to $\partial\Omega$ pointing outwards of Ω
$\mathcal{T}_h$	simplicial regular triangulation of the domain Ω, that is, the set of simplexes T, Definition 2.1.9
$\mathcal{E}_h$, $\mathcal{V}_h$	set of edges or vertices of $\mathcal{T}_h$, respectively
$\mathcal{E}_T$	set of edges of an element $T \in \mathcal{T}_h$
$\mathcal{N}_i^{\mathcal{T}_h}$, $\mathcal{N}_i^{\mathcal{E}}$	set of elements or edges sharing a common vertex v_i; page 102
n_T, n_E, n_v	number of elements, edges and vertices of $\mathcal{T}_h$
h	mesh size of a quasi-uniform triangulation; see Definition 2.1.10
ρ	shape-regularity parameter; see (2.29)
$\mathcal{C}(\Omega)$	space of continuous functions defined on Ω
$\mathcal{P}_k(D)$	space of polynomials on the domain of order $\leq k$, see page 20
S_h	the space of piecewise constant functions on $\mathcal{T}_h$; see (2.30)
V_h^k	the space of piecewise polynomials of order $\leq k$ on $\mathcal{T}_h$; see (2.31)
$\boldsymbol{V}_{h,0}^{RT}$	space of Raviart-Thomas functions on $\mathcal{T}_h$ of order 0; see (2.32)
φ_i, $\boldsymbol{\varphi}_E^{RT}$	nodal basis functions of V_h^1 and $\boldsymbol{V}_{h,0}^{RT}$;pages 21, 102
$N_E^{RT}(.)$	degree of freedom of $\boldsymbol{V}_{h,0}^{RT}$ corresponding to an edge E; see page 21
$\Pi^{RT}(.)$	projection from $H(\mathrm{div};\Omega)$ to $\boldsymbol{V}_{h,0}^{RT}$; see equation (5.1)
$\boldsymbol{n}_E$	globally predefined unit normal vector of $E \in \mathcal{E}_h$
$\boldsymbol{n}_{E,T}$	unit normal vector of $E \in \mathcal{E}_h$, pointing outwards of $T \in \mathcal{T}_h$, $E \in \overline{T}$
$\lambda_{\min}(A)$, $\lambda_{\max}(A)$	minimal/maximal eigenvalue of a symmetric operator $A : V \to V$; then $\lambda_{\min}(A) := \inf_{v\in V} \frac{(Av,v)}{\|v\|}$ and $\lambda_{\max}(A) := \sup_{v\in V} \frac{(Av,v)}{\|v\|}$
$\kappa(A)$	condition number of an SPD matrix A, i.e., $\kappa(A) = \frac{\lambda_{\max}(A)}{\lambda_{\min}(A)}$
$\boldsymbol{\sigma}$, ε, $\boldsymbol{u}$	stress, strain and displacements in linear elasticity; Section 2.3
λ, μ, E, ν	material parameters for isotropic, homogeneous materials; page 25
P_0	L^2-projection onto constants on $T \in \mathcal{T}_h$ or $E \in \mathcal{E}_h$; see (2.54), (2.59)
T_i, P_i	projections onto a subspace $V_i \subset V$ according to (3.5) and (3.15)
M, S	smoother of (A)MG with error propagation operator S; page 43
P	prolongation operator of (A)MG; Section 3.2
$\mathcal{V}$, $\mathcal{E}$	set of algebraic vertices and edges in AMGm; see Definition 4.2.1
s_{ij}	strength of connectivity between two algebraic vertices; Definition 4.3.1
$\mathcal{S}_i$	set of strongly connected neighbors of vertex v_i; page 76
$\mathcal{V}_f$, $\mathcal{V}_c$	set of coarse and fine grid vertices; page 76
$\mathrm{supp}(v)$	support of a function v; $\mathrm{supp}(v) := \{\boldsymbol{x} : v(\boldsymbol{x}) \neq 0\}$
$\mathrm{span}\{\boldsymbol{v}_i\}$	span of the vectors/functions $\boldsymbol{v}_i$; $\mathrm{span}\{\boldsymbol{v}_i\} := \{\sum_{i=1}^n \alpha_i \boldsymbol{v}_i : \alpha_i \in \mathbb{R}\}$

$a \lesssim b$	if there exists $c > 0$ such that $a \leq c\,b$; c being independent of certain parameters, or of the choice of a and b, if $a \in V_1$ and $b \in V_2$
$a \gtrsim b$	complementary to $a \lesssim b$
$a \approx b$	if $a \lesssim b$ and $a \gtrsim b$

For the following considerations, let A be a self-adjoint positive semidefinite operator on $\boldsymbol{V} = \boldsymbol{V}_1 \times \boldsymbol{V}_2$ for some Hilbert spaces $\boldsymbol{V}_1$ and $\boldsymbol{V}_2$. It can be represented in matrix form

$$A = \begin{pmatrix} A_{11} & A_{12} \\ A_{21} & A_{22} \end{pmatrix} . \tag{1.1}$$

Then, the CBS constant $\gamma(A)$ is defined as the minimal constant c for which the *Cauchy-Bunyakovsky-Schwarz inequality* holds

$$|(A_{12}\boldsymbol{v}_2\,,\,\boldsymbol{v}_1)_{\boldsymbol{V}_1}|^2 \leq c^2 (A_{11}\boldsymbol{v}_1\,,\,\boldsymbol{v}_1)_{\boldsymbol{V}_1} (A_{22}\boldsymbol{v}_2\,,\,\boldsymbol{v}_2)_{\boldsymbol{V}_2} \qquad \forall \boldsymbol{v}_1 \in \boldsymbol{V}_1 \forall \boldsymbol{v}_2 \in \boldsymbol{V}_2\,. \tag{1.2}$$

It can be shown, that $\gamma(A) \in [0,\,1]$ (see [Axe94, KM09]). Additionally, if A_{22} is invertible we define the *Schur complement* of A with respect to $\boldsymbol{V}_1$ as

$$S_{11} := A_{11} - A_{12} A_{22}^{-1} A_{21}\,. \tag{1.3}$$

If $\gamma(A) < 1$ is satisfied, $\gamma(A)$ can be alternatively obtained from

$$1 - \gamma(A)^2 = \inf_{\boldsymbol{v}_1 \in \boldsymbol{V}_1} \frac{(S_{11}\boldsymbol{v}_1\,,\,\boldsymbol{v}_1)_{\boldsymbol{V}_1}}{(A_{11}\boldsymbol{v}_1\,,\,\boldsymbol{v}_1)_{\boldsymbol{V}_1}}\,, \tag{1.4}$$

which is shown in [Axe94]. If A_{11} is invertible as well, we can exchange the indices in (1.4).

Chapter 2

Problem setting

The main focus of this work lies in the construction of efficient preconditioners for the linear system of equations arising from finite element discretizations of the steady state equations of linear elasticity.

Therefore, let us first introduce some preliminary tools such as Sobolev spaces and inequalities on Sobolev spaces, which are needed to prove existence and uniqueness of solutions of variational problems being equivalent to certain PDEs. Since we use the finite element method (FEM) to obtain approximations of the solution of variational problems, we also address the basics of FEM. Then, we introduce a scalar elliptic model problem for which the basic properties of FEM discretizations are shown in order to address the general difficulties. In Section 2.3 we formulate the governing equations of linear elasticity. Afterwards, in Section 2.4 we discuss some possible variational formulations of these equations. All of them are solely in the primal variables $\boldsymbol{u}$ representing the displacement from the reference configuration. The advantage of those formulations are, that, after discretization, we are dealing with much less degrees of freedom (DOFs) than for discretizations of mixed variational formulations. Firstly, we show the classical formulation which is then discretized by continuous piecewise linear finite elements. This variational problem suffers from so-called *locking* as the material becomes almost incompressible. In order to circumvent this effect we introduce two stable formulations. The first one uses a technique called *reduced integration*, where the div-part is integrated by less accurate numerical integration techniques. This leads to optimal order error approximations. The third discretization scheme uses discontinuous finite elements. Using proper penalty terms, this formulation can be shown to be robust with respect to the material parameters as well. Finally, in order to provide an almost complete discussion on the possibilities to pose the equations of elasticity variationally, we shortly address mixed formulations of the governing equations.

2.1 Preliminaries

2.1.1 Function spaces

The variational formulations of partial differential equations (PDEs), which are the basis for the finite element method, are posed in *Sobolev* spaces. Let us assume we are given a bounded and connected Lipschitz domain $\Omega \subset \mathbb{R}^d$ with $d = 2, 3$ (see [Gri85] for the definition of Lipschitz domains) and we define

$$L^2(\Omega) := \left\{ v : \Omega \to \mathbb{R} \; : \; \int_\Omega |v|^2 \, \mathrm{d}\boldsymbol{x} < \infty \right\}, \tag{2.1}$$

the space of square-integrable functions. It is equipped with an inner product, which induces a norm,

$$(u\,,\,v)_{L^2(\Omega)} := (u\,,\,v)_{0,\Omega} := \int_\Omega u\,v\,\mathrm{d}\boldsymbol{x}\,, \qquad \|u\|_{L^2(\Omega)} := \sqrt{(u,u)_{L^2(\Omega)}}\,. \tag{2.2}$$

If it follows from the context that we are dealing with the domain Ω we may replace the subscripts "$L^2(\Omega)$" or "$0,\Omega$" by a subscript "0". In the vector-valued case the inner product is defined by

$$(\boldsymbol{u}\,,\,\boldsymbol{v})_0 := \int_\Omega \boldsymbol{u}\cdot\boldsymbol{v}\,\mathrm{d}\boldsymbol{x} = \sum_{i=1}^{d} \int_\Omega u_i v_i \,\mathrm{d}\boldsymbol{x}\,.$$

The definition of the spaces $H^1(\Omega)$ and $H(\mathrm{div};\Omega)$ requires the notion of *generalized* or *weak derivatives*, which we now introduce.

Definition 2.1.1. *A distribution $w \in \mathcal{D}'(\Omega)$ is called* weak derivative *in direction i of a distribution $z \in \mathcal{D}'(\Omega)$ if*

$$\int_\Omega w\,\varphi\,d\boldsymbol{x} = -\int_\Omega z\,\frac{\partial\varphi}{\partial x_i}\,d\boldsymbol{x} \qquad \forall \varphi \in C_0^\infty(\Omega)\,. \tag{2.3}$$

In a similar manner one can define weak gradient "∇", and divergence "div" operators, and then introduce

$$H(D;\Omega) \;:=\; \{\boldsymbol{v} \in [L^2(\Omega)]^k \,:\, Dv \in [L^2(\Omega)]^l\}\,, \tag{2.4}$$

where k and l depend on the differential operator D, e.g., for $D = \nabla$ the constants are $k = 1$ and $l = d$. Specifically, we denote $H^1(\Omega) := H(\nabla;\Omega)$. These spaces are equipped with the inner products

$$(u\,,\,v)_{H^1(\Omega)} \;:=\; (u\,,\,v)_0 + (\nabla u\,,\,\nabla v)_0\,, \tag{2.5}$$

$$(\boldsymbol{u}\,,\,\boldsymbol{v})_{H(\mathrm{div};\Omega)} \;:=\; (\boldsymbol{u}\,,\,\boldsymbol{v})_0 + (\mathrm{div}\,\boldsymbol{u}\,,\,\mathrm{div}\,\boldsymbol{v})_0\,. \tag{2.6}$$

Similar as in the L^2-case we may replace the subscript "$H^1(\Omega)$" by the subscript "$1,\Omega$" or simply by "1". Generally, functions $u \in H^1(\Omega)$ are not defined pointwise on the boundary. Nevertheless, we introduce boundary values of those functions in the sense of traces (cf. [Cia78, Gri85, Bra01]) and define

$$H_0^1(\Omega) := \{u \in H^1(\Omega) \,:\, u|_{\partial\Omega} = 0\}\,, \tag{2.7}$$

or for a subset $\Gamma_D \subset \partial\Omega$ of the boundary

$$H_{0,\Gamma_D}^1(\Omega) \;:=\; \{u \in H^1(\Omega) \,:\, u|_{\Gamma_D} = 0\}\,, \tag{2.8}$$

$$H_{u_D,\Gamma_D}^1(\Omega) \;:=\; \{u \in H^1(\Omega) \,:\, u|_{\Gamma_D} = u_D\}\,. \tag{2.9}$$

If $\mathrm{meas}(\Gamma_D) \neq 0$ then the H^1-norm on $H_0^1(\Omega)$ and $H_{0,\Gamma_D}^1(\Omega)$ is equivalent to the H^1-seminorm given by

$$|u|_{H^1(\Omega)} := (\nabla \boldsymbol{u}\,,\,\nabla \boldsymbol{v})_0\,, \tag{2.10}$$

which is due to Friedrich's inequality (Theorem 2.1.5 presented in the next subsection). For the equations of linear elasticity the space of *rigid body modes* $\boldsymbol{V}^{\mathrm{RBM}}$ is important. The space consists of all translations and rotations, that is

$$\boldsymbol{V}^{\mathrm{RBM}} := \begin{cases} \{\boldsymbol{v} \,:\, \boldsymbol{v} = (a_1 + by,\, a_2 - bx)^t \quad a_1,\, a_2,\, b \in \mathbb{R}\} & \text{for } d = 2\,, \\ \{\boldsymbol{v} \,:\, \boldsymbol{v} = \boldsymbol{a} + \boldsymbol{b} \times \boldsymbol{x},\;\; \boldsymbol{a},\, \boldsymbol{b} \in \mathbb{R}^3\} & \text{for } d = 3\,. \end{cases} \tag{2.11}$$

Further, we define the space $\hat{\boldsymbol{H}}^1(\Omega)$ of H^1-functions complementary to $\boldsymbol{V}_H^{\mathrm{RBM}}$, i.e.,

$$\hat{\boldsymbol{H}}^1(\Omega) := \{\boldsymbol{v} \in [H^1(\Omega)]^d \,:\, \int_\Omega \boldsymbol{v}\,\mathrm{d}\boldsymbol{x} = \boldsymbol{0} \;\text{ and }\; \int_\Omega \nabla \times \boldsymbol{v}\,\mathrm{d}\boldsymbol{x} = \boldsymbol{0}\}\,, \tag{2.12}$$

which is needed for the variational form of the equations of linear elasticity in the case of pure traction boundary conditions (see subsections 2.4.1 and 2.4.2). For $d = 2$ the term $\nabla \times \boldsymbol{v}$ is a scalar while for $d = 3$ it is a 3-dimensional vector. In the following, we prove that $[H^1(\Omega)]^d$ can be uniquely decomposed into $\hat{\boldsymbol{H}}^1(\Omega)$ and $\boldsymbol{V}^{\mathrm{RBM}}$.

Lemma 2.1.2. *We have that* $[H^1(\Omega)]^d = \hat{\boldsymbol{H}}^1(\Omega) \oplus \boldsymbol{V}^{RBM}$, $d = 2,\, 3$.

Proof. Note that for any $\boldsymbol{v} \in \boldsymbol{V}^{\text{RBM}}$ the condition $\boldsymbol{v} \in \hat{\boldsymbol{H}}^1(\Omega)$ holds only if $\boldsymbol{v} = \mathbf{0}$. First, let us consider the case $d = 2$. We introduce the projections $P_c : [H^1(\Omega)]^2 \to \boldsymbol{V}^{\text{RBM}}$ and $P_r : [H^1(\Omega)]^2 \to \boldsymbol{V}^{\text{RBM}}$ via the following relations

$$P_c(\boldsymbol{v}) := \frac{1}{|\Omega|} \int_\Omega \boldsymbol{v} \, \mathrm{d}\boldsymbol{x} \,, \tag{2.13}$$

$$P_r(\boldsymbol{v}) := \frac{-1}{2|\Omega|} \int_\Omega \nabla \times \boldsymbol{v} \, \mathrm{d}\boldsymbol{x} \begin{pmatrix} y \\ -x \end{pmatrix} . \tag{2.14}$$

It can be easily seen that $P_c(.)$ and $P_r(.)$ are indeed projections, i.e., they are idempotent. Now, let us define the total projection $P_{\text{RBM}} : [H^1(\Omega)]^2 \to \boldsymbol{V}^{\text{RBM}}$ through

$$P_{\text{RBM}}(\boldsymbol{v}) := P_r(\boldsymbol{v}) + P_c(\boldsymbol{v} - P_r(\boldsymbol{v})) \,. \tag{2.15}$$

Note that $P_c(P_{\text{RBM}}(\boldsymbol{v})) = P_c(\boldsymbol{v})$ and $P_r(P_{\text{RBM}}(\boldsymbol{v})) = P_r(\boldsymbol{v})$. Now, every $\boldsymbol{v} \in \boldsymbol{V}$ can be decomposed as $\boldsymbol{v} = \boldsymbol{v} - P_{\text{RBM}}(\boldsymbol{v}) + P_{\text{RBM}}(\boldsymbol{v}) = \hat{\boldsymbol{v}} + P_{\text{RBM}}(\boldsymbol{v})$ with $\hat{\boldsymbol{v}} := \boldsymbol{v} - P_{\text{RBM}}(\boldsymbol{v}) \in \hat{\boldsymbol{H}}^1(\Omega)$.

The construction of the projections P_c and P_r can be easily extended for $d = 3$. Thereby, P_r is set up from 3 single projections onto $\boldsymbol{r}_1 = (0, z, -y)^T$, $\boldsymbol{r}_2 = (-z, 0, x)^T$ and $\boldsymbol{r}_3 = (y, -x, 0)^T$ with a proper scaling. That is, $P_r(\boldsymbol{v}) := \sum_{i=1}^3 P_{r_i}(\boldsymbol{v})$ with

$$P_{r_i}(\boldsymbol{v}) := \frac{-1}{2|\Omega|} \int_\Omega (\nabla \times \boldsymbol{v})_i \, \mathrm{d}\boldsymbol{x} \, \boldsymbol{r}_i \,.$$

Hence, this statement is valid for $d = 3$ too. □

Moreover, the space

$$H_S(\Omega) := \{\boldsymbol{\tau} \in [L^2(\Omega)]^{d \times d} \, : \, \boldsymbol{\tau}^T = \boldsymbol{\tau} \text{ a.e.}\} \tag{2.16}$$

is needed in Subsection 2.4.4 for a mixed variational formulation of the equations of linear elasticity.

2.1.2 Preliminary results (inequalities)

The following inequalities are useful in proving solvability of the system of PDEs. The proofs of the first three theorems can be found in [TW05], and Theorems 2.1.6 and 2.1.7 are proven in [BS07].

Theorem 2.1.3 (Poincarè inequality). *Let $u \in H^1(\Omega)$. Then there exist c_1 and c_2, depending* *only on Ω, such that*

$$\|u\|^2_{L^2(\Omega)} \leq c_1 \, |u|^2_{H^1(\Omega)} + c_2 \left(\int_\Omega u \, d\boldsymbol{x} \right)^2 . \tag{2.17}$$

From the previous theorem we obtain by scaling arguments a dependence of the constant c_1 on the diameter of the domain Ω.

Corollary 2.1.4. *Let Ω be a Lipschitz domain with diameter H. Then, there exists a constant $\hat{c}_1$, that depends only on the shape of Ω but not on the size, such that*

$$\|u\|^2_{L^2(\Omega)} \leq \hat{c}_1 H^2 \, |u|^2_{H^1(\Omega)} , \tag{2.18}$$

for $u \in H^1(\Omega)$ with vanishing mean value on Ω.

Friedrich's inequality yields the spectral equivalence of the H^1-seminorm to the H^1-norm on H^1_{0,Γ_D}.

Theorem 2.1.5 (Friedrich's inequality). *Let $\Gamma \subset \partial\Omega$ with* $\text{meas}(\Gamma) > 0$. *Then there exists a $c_F > 0$ depending only on Ω and Γ such that for all $u \in H^1(\Omega)$*

$$\|u\|^2_{H^1(\Omega)} \leq c_F \, |u|^2_{H^1(\Omega)} + c_F \, \|u\|^2_{L^2(\Gamma)} . \tag{2.19}$$

The last type of inequality we need in this book is *Korn's inequality.* It states that the symmetrized gradient $\boldsymbol{\varepsilon}(\boldsymbol{u}) := \frac{1}{2}[\nabla \boldsymbol{u} + (\nabla \boldsymbol{u})^t]$ bounds the H^1-norm from above under certain assumptions. Note that the rigid modes span the kernel of $\boldsymbol{\varepsilon}(.)$. That is, we have $\boldsymbol{\varepsilon}(\boldsymbol{v}) = \boldsymbol{0}$ for all $\boldsymbol{v} \in \boldsymbol{V}^{\text{RBM}}$. For our purposes we need two different versions of the theorem (cf. [BS07]).

Theorem 2.1.6. *There exists a positive constant c_K such that*

$$\|\boldsymbol{\varepsilon}(\boldsymbol{u})\|_{L^2(\Omega)} \geq c_K \, \|\boldsymbol{u}\|_{H^1(\Omega)} \qquad \forall \boldsymbol{u} \in \hat{\boldsymbol{H}}^1(\Omega) . \tag{2.20}$$

If we are dealing with mixed or pure displacement boundary conditions we need another version of Korn's inequality which takes care of the rigid modes, especially the rotations, by a boundary part with zero trace.

Theorem 2.1.7. *Let $\Gamma_D \subset \partial\Omega$ with* $\text{meas}(\Gamma_D) > 0$. *There exists a positive constant c_K such that*

$$\|\boldsymbol{\varepsilon}(\boldsymbol{u})\|_{L^2(\Omega)} \geq c_K \, \|\boldsymbol{u}\|_{H^1(\Omega)} \qquad \forall \boldsymbol{u} \in [H^1_{0,\Gamma_D}(\Omega)]^d . \tag{2.21}$$

For a further discussion on Korn's inequality see [DM04, KO89]. Therein, it is mentioned that in the case $\Gamma_D = \partial\Omega$ $c_K = \sqrt{2}$ for any open set Ω. In [DM04] the inequality is discussed for Jones domains, which are more general than Lipschitz domains (cf. [Jon81]). It is stated that the constant c_K in all cases does only depend on the shape of the domain Ω.

Lemma 2.1.8. *Let V be a Banach space. Further, for finite $J \in \mathbb{N}$, $J \geq 2$ let V_i, $i = 1, \ldots, J$, be closed subspaces of V such that*

$$V = \sum_{i=1}^{J} V_i \,. \tag{2.22}$$

Then, there exists a $c > 0$ such that for any $v \in V$ there exist $v_i \in V_i$, $i = 1, \ldots, J$ with

$$c \sum_{i=1}^{J} \|v_i\| \leq \|v\| \,. \tag{2.23}$$

Proof. Let us define the Banach space $\tilde{V} = V_1 \times V_2 \times \ldots \times V_J$ with norm $\|\tilde{v}\|_{\tilde{V}} := \sum_{i=1}^{J} \|v_i\|$ and the mapping $T : \tilde{V} \to V$ defined by $T(\tilde{v}) = \sum_{i=1}^{J} v_i$. Condition (2.22) implies the surjectivity of T. Hence, due to the *open mapping theorem* (cf. [RS80, p. 82]) T is *open*. Equivalently, this means that for $r > 0$ and $\tilde{U}_r := \{\tilde{v} \in \tilde{V} \,:\, \|\tilde{v}\|_{\tilde{V}} < r\}$ it holds that there exists $c > 0$ with $U_c \subset T(\tilde{U}_1)$ for $U_c := \{v \in V \,:\, \|v\| < c\}$. Since T is linear, for any $v \in U_1$ we find that $v \in T(\tilde{U}_{\frac{1}{c}})$. Finally,

$$\sup_{\|v\|<1} \inf_{\tilde{v} : v = T(\tilde{v})} \|\tilde{v}\|_{\tilde{V}} = \sup_{\|v\|=1} \inf_{\tilde{v} : v = T(\tilde{v})} \|\tilde{v}\|_{\tilde{V}} < \frac{1}{c} \,,$$

and hence, through scaling, we obtain

$$c \inf_{\tilde{v} : v = T(\tilde{v})} \|\tilde{v}\|_{\tilde{V}} = c \inf_{\tilde{v} : v = T(\tilde{v})} \sum_{i=1}^{J} \|v_i\| < \|v\| \qquad \forall v \in V \,,$$

which implies the statement. □

2.1.3 Finite element discretization

In this book we are concerned with the solution of linear systems of equations arising from the finite element discretization of partial differential equations (PDEs). Therefore a short summary on the Galerkin finite element method (FEM) is presented. More details on the method and related issues can be found in various textbooks such as [Cia78, BS07, Bra01]. The details on mixed and hybrid finite element formulations are summarized for instance in [BF91].

The FEM deals with variational formulations of boundary value problems. For this sake the PDE is multiplied by a suitable test function (of a certain Sobolev space). Then we integrate over the whole domain Ω and apply properly the integration by parts rule.

Viewed in an abstract setting we do have a Hilbert space H or a closed subspace $V \subset H$ equipped with an inner product $(.,.)$. Further we are given a bilinear form $a(.,.) : H \times H \to \mathbb{R}$ and a linear functional $L \in H'$ and consider the following problem:

Find $u \in V$ such that

$$a(u, v) = L(v) \qquad \forall v \in V\,. \tag{2.24}$$

It is well known that this problem has a unique solution if $a(.,.)$ is bounded and coercive, i.e., if there exist $\bar{c}$, $\underline{c} > 0$ such that

$$|a(u, v)| \le \bar{c}\, \|u\|\, \|v\| \qquad \forall u, v \in V\,, \tag{2.25}$$

$$|a(u, u)| \ge \underline{c}\, \|u\|^2 \qquad \forall u \in V\,, \tag{2.26}$$

by the Theorem of Lax-Milgram.

In general the problem (2.24) cannot be solved exactly. For this reason the problem is approximated. Usually the Hilbert space V is replaced by a finite dimensional subspace V_h. The subscript h denotes the discretization parameter. In our case H is either of the Sobolev spaces $H^1(\Omega)$ or $H(\mathrm{div};\Omega)$.

For the construction of proper subspaces of finite dimension we subdivide the domain Ω into triangles or quadrangles for $d = 2$. In the three-dimensional case the triangulation is based on tetrahedra or hexagons. For the sake of simplicity we assume a polygonal or polyhedral Lipschitz domain $\Omega \subset \mathbb{R}^d$, $d = 2, 3$, which is sufficient for our considerations.

We assume the triangulation $\mathcal{T}$ to be *regular* which is defined as

Definition 2.1.9 (regular triangulation). *A triangulation $\mathcal{T} = \{T\}$ is called* regular *if it is a non-overlapping decomposition of the domain Ω into open elements T of simple geometry, i.e., triangles or quadrangles in 2D and tetrahedra or hexahedra in 3D. Further $\mathcal{T}$ has to fulfill the following properties:*

1. *the elements $T, \tilde{T} \in \mathcal{T}$ have to be non-overlapping, i.e., $T \cap \tilde{T} = \emptyset$ if $T \neq \tilde{T}$,*
2. *$\mathcal{T}$ has to cover the whole domain $\overline{\Omega} = \bigcup_{T\in\mathcal{T}} \overline{T}$,*
3. *the intersection of the closure of two neighboring elements $\overline{T}$ and $\overline{\tilde{T}} \in \mathcal{T}$ is either a common face, edge or vertex of both elements.*

Such a triangulation is also referred to as *mesh.* As a measure for the element size we use its diameter $h_T := \operatorname{diam} T$. B_T denotes the largest ball that can be inscribed into T. Additionally, we denote the ratio $\operatorname{diam}(B_T)/h_T$ by ρ_T. Following [BS07] we define

Definition 2.1.10. *Let $\{\mathcal{T}_h\}$, $0 < h \leq 1$ be a family of subdivisions with*

$$h := \max\{h_T \,:\, T \in \mathcal{T}_h\}\,. \tag{2.27}$$

The family is said to be quasi-uniform *if there exists a $\rho > 0$ for all h such that*

$$\min\{\operatorname{diam} B_T \,:\, T \in \mathcal{T}_h\} \geq \rho h\,. \tag{2.28}$$

The family is called shape-regular *if there is a $\rho > 0$ for all $T \in \mathcal{T}_h$ and all h satisfying*

$$\rho_T \geq \rho\,. \tag{2.29}$$

Note that quasi-uniformity implies shape-regularity. In a shape-regular mesh the interior angles of all triangles and tetrahedra are bounded. This means that there is no element T for which the ratio of minimal edge length to the diameter of the element h_T deteriorates. For shape-regular meshes $\mathcal{T}_h$ the subscript h represents the mesh size, that is, $h := \max_{T \in \mathcal{T}_h} h_T$.

For a certain $\mathcal{T}_h$ we define the set of vertices of the elements by $\mathcal{V}_h$. Moreover, $\mathcal{E}_h$ denotes the set of edges of the mesh. If it is clear from the context that we do consider $\mathcal{T}_h$ we may skip the subscript h for all the mentioned sets.

Following [Cia78] a finite element consists of an element domain (usually $T \in \mathcal{T}$), a finite dimensional space of functions $\{\varphi_j\}$, called *shape functions* and a basis for the dual of the shape functions $\{N_i\}$, that is, the set of *nodal variables* or sometimes the set of *degrees of freedom.* All three components together are called a *finite element.* If we have $N_i(\varphi_j) = \delta_{ij}$ the basis is called *nodal basis.*

In this book we only focus on polynomial shape functions. Furthermore we restrict ourselves to a triangulation consisting only of triangles or tetrahedra and use only nodal basis functions. By $\mathcal{P}_k(D)$ we denote the set of polynomials of degree less or equal to k on a domain D. We need subspaces of $L^2(\Omega)$, $H^1(\Omega)$ and $H(\operatorname{div};\Omega)$. Therefore, let us define the space of piecewise constant functions S_h, the continuous piecewise polynomial functions V_h^k, $k \geq 1$, and the $H(\operatorname{div};\Omega)$-conforming piecewise linear *Raviart-Thomas space* $\mathbf{V}_{h,0}^{RT}$ of order 0 on a triangulation $\mathcal{T}_h$ as follows

$$
\begin{aligned}
S_h &:= \{v \in L^2(\Omega) : v|_T \in \mathcal{P}_0(T) \quad \forall T \in \mathcal{T}_h\}, &(2.30)\\
V_h^k &:= \{v \in \mathcal{C}(\Omega) : v|_T \in \mathcal{P}_k(T) \quad \forall T \in \mathcal{T}_h\}, &(2.31)\\
\boldsymbol{V}_{h,0}^{RT} &:= \{\boldsymbol{v} \in H(\mathrm{div};\Omega) : \boldsymbol{v}|_T = \boldsymbol{a} + b\boldsymbol{x} \quad \forall T \in \mathcal{T}_H,\ \boldsymbol{a} \in \mathbb{R}^d,\ b \in \mathbb{R}\}. &(2.32)
\end{aligned}
$$

Since $\mathcal{C}(\Omega) \subset H^1(\Omega)$ we find that V_h^k is H^1-conforming. For $k = 1$ the piecewise linear functions are defined via their values at the vertices of $\mathcal{T}_h$. For $k > 1$ evaluation points are added properly.

In the case of $\boldsymbol{V}_{h,0}^{RT}$ the degrees of freedom are defined by the normal flux across edges ($d = 2$) or faces ($d = 3$), i.e.,

$$N_E^{RT}(\boldsymbol{v}) := \frac{1}{|E|}\int_E \boldsymbol{v} \cdot \boldsymbol{n}_E \,\mathrm{d}s.$$

Here $\boldsymbol{n}_E$ denotes the unit normal vector of E, which is predefined (globally) on each edge/face. On boundary faces $E \subset \partial\Omega$ we define $\boldsymbol{n}_E$ to be the outward unit vector of the unique element T associated with the face E. This implies that the space $\boldsymbol{V}_{h,0}^{RT}$ consists of piecewise linear functions with continuous normal components in the edge/face midpoints. For more information on Raviart-Thomas FE spaces see [BF91].

Now, let us come back to the abstract setting. In order to discretize the problem (2.24), we are given a finite element space $V_h := \mathrm{span}\{\varphi_i\}$ being the span of nodal basis functions. We formulate the discrete problem as:

Find $u_h \in V_h$ such that

$$a(u_h, v_h) = L(v_h) \qquad \forall v_h \in V_h. \tag{2.33}$$

Existence and uniqueness of the solution u_h follow from the conformity of the finite element space, $V_h \subset V$, and from the results on the original problem (2.24). We apply the isomorphism from V_h to $\mathbb{R}^n$ with $n = |V_h|$ to get $\boldsymbol{u}_h = (u_i)$ by $u_h = \sum_{i=1}^n u_i\varphi_i$ and arrive at the linear system

$$A\boldsymbol{u}_h = \boldsymbol{f}_h \tag{2.34}$$

where $A = (a_{ij}) = (a(\varphi_j, \varphi_i))$ and $\boldsymbol{f}_h = (f_i) = (L(\varphi_i))$.

2.2 Scalar elliptic model problem

The standard example for an elliptic PDE is the diffusion problem

$$-\operatorname{div}(\Lambda\nabla u) = f \qquad \text{in } \Omega \tag{2.35a}$$

$$u = g_D \qquad \text{on } \Gamma_D \tag{2.35b}$$

$$\frac{\partial u}{\partial \boldsymbol{n}} = g_N \qquad \text{on } \Gamma_N\,. \tag{2.35c}$$

Thereby Γ_D and Γ_N are subsets of the boundary $\partial\Omega = \Gamma_D \cup \Gamma_N$. Due to homogenization considerations we may assume without loss of generality that $g_D = 0$ on Γ_D. Λ is a $d\times d$-tensor which generally depends on $\boldsymbol{x}$. The variational form of this problem is given as:
Find $u \in H^1_{0,\Gamma_D}(\Omega)$ such that

$$\int_\Omega (\Lambda(\boldsymbol{x})\nabla u(\boldsymbol{x}))\cdot\nabla v(\boldsymbol{x})\,\mathrm{d}\boldsymbol{x} = \int_\Omega f(\boldsymbol{x})v(\boldsymbol{x})\,\mathrm{d}\boldsymbol{x} + \int_{\Gamma_N} g_N(s)v(s)\,\mathrm{d}s \qquad \forall v\in H^1_{0,\Gamma_D}(\Omega)\,. \tag{2.36}$$

If we have c_1, $c_2 > 0$ such that

$$c_1\boldsymbol{z}^t\boldsymbol{z} \le \boldsymbol{z}^t\Lambda(\boldsymbol{x})\boldsymbol{z} \le c_2\boldsymbol{z}^t\boldsymbol{z} \qquad \forall\boldsymbol{x}\in\Omega\,\forall\boldsymbol{z}\in\mathbb{R}^d\,, \tag{2.37}$$

the equation is coercive on $H^1_{0,\Gamma_D}(\Omega)$ due to Friedrich's inequality (Theorem 2.1.5) with $\underline{c} = c_1 c_F$ and bounded with $\bar{c} = c_2$. Hence, the solution u does exist and it is unique. After a proper triangulation $\mathcal{T}_h$ is constructed, we seek for a discrete solution $u_h \in V_h^k$, which ends up in the linear system

$$A\boldsymbol{u}_h = \boldsymbol{f}_h\,. \tag{2.38}$$

Since the number of unknowns n might become huge, the system (2.38) cannot be solved directly. This would require $\mathcal{O}(n^3)$ floating point operations for a naive direct solve of the sparse system. With nested dissection ordering one obtains a complexity of $\mathcal{O}(n^{3/2})$ floating point operations for 2D problems and for $d = 3$ we have $\mathcal{O}(n^2)$ operations when nested dissection is used (cf. [GL81]). This is impractical and hence one needs to apply iterative solution methods. Thereby, the condition number κ of the system matrix A usually plays an crucial role. We find the following spectral bounds for A_h (cf. [Joh87] for $d = 2$) and small k

$$\lambda_{\max}(A_h) \lesssim h^{d-2}\lambda_{\max}(\Lambda) \quad\text{and}\quad \lambda_{\min}(A_h) \gtrsim h^d\Lambda_{\min}(\Lambda)\,. \tag{2.39}$$

Consequently, the condition number $\kappa(A_h)$ is of the order $\mathcal{O}(\frac{\lambda_{\max}(\Lambda)}{h^2\lambda_{\min}(\Lambda)})$ for moderate polynomial degree k. That is, we have for finer and finer meshes a strongly increasing factor h^{-2} and additionally, for jumping or varying coefficients of the matrix Λ, the range of the spectrum of Λ directly influences the condition number. For higher order polynomial approximations k enters the condition number of A_h according to [MP96] with the factor $k^{4(d-1)}$ or with $k^{2(d-1)}$ if diagonal scaling is used. Nevertheless, we focus on $k \leq 3$.

Due to the bad condition number of the matrix A_h we additionally need efficient preconditioners in order to solve the linear problem with a complexity of $\mathcal{O}(n)$. That is, because the convergence of standard iterative procedures such as Richardson, Jacobi or Gauss-Seidel iteration deteriorate when applied directly to such problems.

Another iterative solution method is the conjugate gradient (CG) method introduced in [HS52], see also [Axe94, Vas08, KM09]. Its i-th iterate $\boldsymbol{u}^{(i)}$ fulfills

$$\|\boldsymbol{u}^{(i)} - \boldsymbol{u}\|_{A_h} \leq 2 \left(\frac{\sqrt{\kappa(A_h)} - 1}{\sqrt{\kappa(A_h)} + 1} \right)^i \|\boldsymbol{u}^{(0)} - \boldsymbol{u}\|_{A_h}, \tag{2.40}$$

where $\boldsymbol{u}^{(0)}$ is the initial guess and $\boldsymbol{u}$ is the exact solution. Even though one iteration is of the order $A_h\boldsymbol{u}$, that is $\mathcal{O}(n)$, the convergence factor tends to 1 for increasing n. A remedy to this is to devise a uniform preconditioner B_h of A_h. Then, for the preconditioned CG (PCG) method the convergence estimates are equivalent to (2.40), but now, $\kappa(B_h^{-1}A_h)$ enters. Hence, if B_h is such that $\kappa(B_h^{-1}A_h)$ is uniformly bounded, the PCG solver is of optimal order.

2.3 The equations of linear elasticity

The field of continuum mechanics deals with the deformation of materials and different substances. Here, we are interested in the behavior of solids under certain loads or forces, respectively. In the following, we briefly sketch the needed ingredients and equations which finally yield the equations of linear elasticity. The focus of this work lies on the steady state case and hence all considered variables are time-independent. For a thorough discussion see [MH94] or the nice summary in [Bra01].

Let us assume we are given a body (piece of material) $\Omega \subset \mathbb{R}^d, d = 2, 3$, also called *reference configuration.* Later on, we need Ω to be a smooth and connected domain. Every point $\boldsymbol{X} \in \Omega$ is called material point, whereas $\boldsymbol{x} \in \mathbb{R}^d$ is called spatial point. A mapping $\boldsymbol{\Phi} : \Omega \to \mathbb{R}^d$ which preserves the orientation and which is invertible, i.e.,

$$\det(\nabla\boldsymbol{\Phi}(\boldsymbol{X})) > 0$$

for all $\boldsymbol{X} \in \Omega$ is called *deformation* of Ω. $\boldsymbol{x} = \boldsymbol{\Phi}(\boldsymbol{X})$ denotes the deformed spatial point of the material point $X \in \Omega$. Further, we introduce the *displacement* $\boldsymbol{U}(\boldsymbol{X}) := \boldsymbol{\Phi}(\boldsymbol{X}) - \boldsymbol{X}$ and $\boldsymbol{u}(\boldsymbol{x}) = \boldsymbol{U}(\boldsymbol{X})$. The nonlinear *Green's strain tensor* $\boldsymbol{E}(\boldsymbol{U})$ characterizes the local deformation of the body. It vanishes if the motion is a *rigid body transformation*, given by

$$\boldsymbol{\Phi}(\boldsymbol{X}) = Q\boldsymbol{X} + \boldsymbol{b}$$

with $\boldsymbol{b} \in \mathbb{R}^d$ and with an orthogonal tensor $Q \in \mathbb{R}^{d\times d}$ $(Q^tQ = I)$. Here Q represents the rotations and $\boldsymbol{b}$ the translations. In 2 dimensions we have 3 rigid motions (2 translations and 1 rotation), while in 3D we do have 6 independent rigid body transformations (3 rotations and 3 translations). Under the assumption that we are only dealing with small displacements, the material points $\boldsymbol{X}$ can be replaced by the spatial points $\boldsymbol{x}$, as well as the displacement $\boldsymbol{U}(\boldsymbol{X})$ by $\boldsymbol{u}(\boldsymbol{x})$, and moreover, the Green strain tensor can be replaced by its linearization $\boldsymbol{\varepsilon}(\boldsymbol{u})$, also called the *symmetrized gradient*, which is

$$\boldsymbol{\varepsilon}(\boldsymbol{u})(\boldsymbol{x}) := \frac{1}{2}(\nabla\boldsymbol{u}(\boldsymbol{x}) + [\nabla\boldsymbol{u}(\boldsymbol{x})]^t), \tag{2.41}$$

or equivalently, $\boldsymbol{\varepsilon}(\boldsymbol{u}) = (\varepsilon_{ij}(\boldsymbol{u}))_{i,j=1,\dots,d}$ where $\varepsilon_{ij}(\boldsymbol{u}) = \frac{1}{2}\left(\frac{\partial u_i}{\partial x_j} + \frac{\partial u_j}{\partial x_i}\right)$. The equations of linear elasticity are based on balance equations. Therefore, the *surface force density* is introduced. A basic axiom of continuum mechanics is the balance of momentum. It states that this surface force density (stress) balances out the forces acting on every surface within Ω (gravity and surface forces). Further, it can be shown (cf. [MH94]) that this force density depends linearly on the normal vector of the specific surface. This is referred to as Cauchy's theorem and it implies the existence of a *stress tensor field* $\boldsymbol{\sigma}$.

The balance of momentum yields the steady state *Cauchy's equation of motion*

$$0 = \operatorname{div}\boldsymbol{\sigma} + \boldsymbol{f} \tag{2.42}$$

together with the condition that $\boldsymbol{\sigma}$ is symmetric. Thereby, $\boldsymbol{f}$ designates the body forces acting on Ω.

Finally, *material laws* link the stress tensor $\boldsymbol{\sigma}$ and the linearized strain tensor $\boldsymbol{\varepsilon}$ to each other. We consider only *linear elastic* materials which means that the strain depends linearly on the stress. This is also referred to as *Hooke's law*. It implies the existence of a fourth-order tensor field $C(\boldsymbol{x})$ such that

$$\boldsymbol{\sigma}(\boldsymbol{u})(\boldsymbol{x}) = C(\boldsymbol{x})\boldsymbol{\varepsilon}(\boldsymbol{u})(\boldsymbol{x}) \qquad \text{in } \Omega. \tag{2.43}$$

Moreover, if C does not depend on $\boldsymbol{x}$ the material is called *homogeneous*. A further material property is *isotropy*. An isotropic material's properties are direction-independent, i.e., invariant with respect to rotations. Otherwise the material is called anisotropic. For isotropic materials the relation between stress and strain simplifies to

$$\boldsymbol{\sigma} = \frac{E}{1+\nu}\left(\frac{\nu}{1-2\nu}\operatorname{tr}(\boldsymbol{\varepsilon}(\boldsymbol{u}))I + \boldsymbol{\varepsilon}(\boldsymbol{u})\right), \tag{2.44}$$

where the scalar E denotes the *Young's modulus* and ν is the *Poisson ratio.* Their physically reasonable values are

$$E > 0 \qquad \text{and} \qquad 0 < \nu < 1/2\,.$$

People sometimes use the *Lamé constants* λ and μ (also called *shear modulus*) to express the stress-strain relation. With the transformation

$$\lambda = \frac{E\nu}{(1-2\nu)(1+\nu)} \qquad \text{and} \qquad \mu = \frac{E}{2(1+\nu)} \tag{2.45}$$

we arrive at

$$\boldsymbol{\sigma} = \lambda \operatorname{tr}(\boldsymbol{\varepsilon}(\boldsymbol{u}))I + 2\mu\boldsymbol{\varepsilon}(\boldsymbol{u})\,. \tag{2.46}$$

For $\nu < 1/2$ or $\lambda < \infty$, respectively, the tensor C is invertible. Since we know that $\boldsymbol{\sigma}$ and $\boldsymbol{\varepsilon}$, respectively, are symmetric we can write the tensors $\boldsymbol{\sigma}$ and $\boldsymbol{\varepsilon}$ in vector form of length 6 ($d = 3$) or 3 ($d = 2$). Let us assume the following arrangement $\boldsymbol{\sigma} = (\sigma_{11}, \sigma_{22}, \sigma_{33}, \sigma_{12}, \sigma_{13}, \sigma_{23})^t$ and $\boldsymbol{\varepsilon} = (\varepsilon_{11}, \varepsilon_{22}, \varepsilon_{33}, 2\varepsilon_{12}, 2\varepsilon_{13}, 2\varepsilon_{23})^t$ in 3D. In 2D we have $\boldsymbol{\sigma} = (\sigma_{11}, \sigma_{22}, \sigma_{12})^t$ and $\boldsymbol{\varepsilon} = (\varepsilon_{11}, \varepsilon_{22}, 2\varepsilon_{12})^t$. Then we can write for isotropic, homogeneous materials the tensor C as second-order tensor. It is given by its inverse C_{iso}^{-1}, called *compliance tensor*,

$$C_{\text{iso}}^{-1} := \begin{pmatrix} 1/E & -\nu/E & -\nu/E & 0 & 0 & 0 \\ -\nu/E & 1/E & -\nu/E & 0 & 0 & 0 \\ -\nu/E & -\nu/E & 1/E & 0 & 0 & 0 \\ 0 & 0 & 0 & 1/\mu & 0 & 0 \\ 0 & 0 & 0 & 0 & 1/\mu & 0 \\ 0 & 0 & 0 & 0 & 0 & 1/\mu \end{pmatrix}, \tag{2.47}$$

The subscript "iso" indicates the isotropic behavior.

In the case of an anisotropic body, which is the most general constitutive law for elastic materials, the tensor C consists of 36 unknowns which is due to the symmetry of the tensors $\boldsymbol{\sigma}$ and $\boldsymbol{\varepsilon}$. If the material exhibits certain axes of symmetry, then C can be further simplified. An *orthotropic* material behaves differently along certain orthogonal directions. Usually we align them with the coordinate system. In that case the compliance tensor reads as

$$C_{\text{ortho}}^{-1} := \begin{pmatrix} 1/E_1 & -\nu_{12}/E_1 & -\nu_{13}/E_1 & 0 & 0 & 0 \\ -\nu_{12}/E_1 & 1/E_2 & -\nu_{23}/E_2 & 0 & 0 & 0 \\ -\nu_{13}/E_1 & -\nu_{23}/E_2 & 1/E_3 & 0 & 0 & 0 \\ 0 & 0 & 0 & 1/\mu_{23} & 0 & 0 \\ 0 & 0 & 0 & 0 & 1/\mu_{13} & 0 \\ 0 & 0 & 0 & 0 & 0 & 1/\mu_{12} \end{pmatrix} . \tag{2.48}$$

E_i is the Young's modulus in the x_i-direction, μ_{ij} is the shear modulus in the x_i-x_j-plane and ν_{ij} is the major Poisson ratio. This constitutive law is determined by 9 unknowns.

Finally, in order to have a system with a unique solution we have to prescribe certain boundary conditions. Then we end up with a *boundary value problem.* It is intuitionally apparent that when we fix some part of the boundary, that the deformation of a body under certain loads is unique. This confirms the theory, because if we fix the displacement on some part of the boundary with positive measure the rigid body transformations are excluded. On the other hand for purely Neumann boundary conditions, i.e., for fixed forces on the whole boundary the solution is unique up to rigid modes, which do not influence the stress and the strain, respectively.

Let us denote by $\Gamma := \partial\Omega$ the boundary of the domain $\Omega \subset \mathbb{R}^d$, $d = 2, 3$. We subdivide Γ into two disjoint sets Γ_D and Γ_N. They fulfill $\Gamma = \Gamma_D \cup \Gamma_N$. On Γ_D we prescribe *Dirichlet boundary conditions* where we fix the displacement $\boldsymbol{u}$ to some given value $\boldsymbol{u}_D$ on the surface

$$\boldsymbol{u} = \boldsymbol{u}_D \qquad \text{on } \Gamma_D .$$

On the remaining part Γ_N we set so-called *Neumann boundary conditions.* Thereby we impose the surface traction on the boundary, i.e.,

$$\boldsymbol{\sigma} \cdot \boldsymbol{n} = \boldsymbol{t}_N \qquad \text{on } \Gamma_N .$$

One might consider additionally pressure boundary conditions which are a special case of the Neumann boundary case.

2.4 Variational Formulations

In this section we address three different variational formulations of the equations of linear elasticity (2.42), for which we discuss suitable preconditioners in the following chapters. The first one is the standard pure displacement formulation for which we observe the effect of *locking*. To overcome the problem of volume locking we introduce a bilinear form which uses an approximative integration, called *reduced integration*. A discontinuous Galerkin (DG) formulation of the governing equations is another possibility to have optimal error estimates independently of the Poisson ratio. Finally, to complete the picture we discuss some mixed formulations.
As discussed for the scalar model problem we assume homogeneous Dirichlet boundary conditions $\boldsymbol{u}_D = \boldsymbol{0}$ on Γ_D.

2.4.1 Pure displacement formulation

The first variational formulation is the simplest (cf. [Bra01, Section 6.3]). We multiply (2.42) by a suitable test function and apply the integration by parts formula. Using the symmetry of $\varepsilon(\boldsymbol{u})$ we arrive at the following variational problem for meas$(\Gamma_D) > 0$:

Find $\boldsymbol{u} \in [H^1_{0,\Gamma_D}(\Omega)]^d$ such that (2.24) holds with

$$a(\boldsymbol{u}, \boldsymbol{v}) := \int_\Omega C\varepsilon(\boldsymbol{u}) : \varepsilon(\boldsymbol{v})\,\mathrm{d}\boldsymbol{x} = (C\varepsilon(\boldsymbol{u})\,,\,\varepsilon(\boldsymbol{v}))_0\,, \tag{2.49}$$

$$L(\boldsymbol{v}) = (\boldsymbol{f}\,,\,\boldsymbol{v})_0 - \int_{\Gamma_N} \boldsymbol{t}_N^T \boldsymbol{v}\,\mathrm{d}s\,, \tag{2.50}$$

and the test space $V := [H^1_{0,\Gamma_D}(\Omega)]^d$. If $\nu < 1/2$ and $E > 0$ on the whole domain Ω we find due to Korn's inequality (Theorem 2.1.7) that $a(.,.)$ is coercive on V. Moreover, the bilinear form is bounded and hence we have existence and uniqueness of the solution.

In the pure traction boundary case we have to alter V and choose $V := \hat{\boldsymbol{H}}^1(\Omega)$ and obtain the coercivity of $a(.,.)$ due to Theorem 2.1.6. Existence and uniqueness of the solution follow if the right-hand side $L(.)$ fulfills the compatibility condition

$$L(\boldsymbol{r}) = 0 \qquad \forall \boldsymbol{v} \in \boldsymbol{V}^{\mathrm{RBM}}\,, \tag{2.51}$$

which is due to Fredholm's theorem (see e.g. [Fal91, BS07, KO89] for a proof).

For the discretization we choose a triangulation $\mathcal{T}_h$ of the domain and use the piecewise polynomial space $[V_h^k]^d$. Due to the conformity we do have a unique solution $\boldsymbol{u}_h$.

Having a closer look at the coercivity and boundedness of (2.49) we observe for an isotropic material $\underline{c} = c_K \lambda_{\min}(C) = c_K \frac{E}{1+\nu}$ and $\bar{c} = \lambda_{\max}(C) = \frac{E}{1-2\nu}$. Hence, we find a deterioration of the discretization which results in a ill-conditioned problem as the material parameter ν tends to 1/2. People in finite element community call this phenomenon *locking*, or in this case *volume locking*.

Locking has been theoretically considered in [Arn81, BS92b, BS92a] and nicely summarized in [Bra01]. The basic message is that we observe locking if the error approximation

$$\|\boldsymbol{u} - \boldsymbol{u}_h\|_1 = \mathcal{O}(h)$$

is not uniform with respect to a small parameter $t \to 0$, which is in our case $\nu \to 1/2$ or equivalently $\lambda \to \infty$.

Contrary to the locking due to the Poisson ratio there is *shear locking*. This occurs if the shape-regularity deteriorates, i.e. if ρ in (2.29) tends to zero. Especially in thin structures such as beams or shells the triangulation might become anisotropic from which the accuracy suffers . Especially the constant in Korn's inequality tends to zero for meshes that are not shape-regular. Such examples have been considered for instance in [Sch99b].

The discretization of (2.49) with piecewise polynomial functions suffers from volume locking for a degree $k \leq 3$. For $k \geq 4$ one obtains optimal error estimates on quasi-uniform meshes as shown in [SV85]. In Figure 2.1 the the effect of locking is demonstrated for $k = 1, 2, 3$ and $k = 4$.

2.4.2 Reduced Integration

In the following we provide a bilinear form for which one can show optimal order error estimates uniformly in the Poisson ratio ν for continuous piecewise linear finite elements. This approach uses the so-call *reduced integration technique*, which goes back to [ESGB82] and [HLB79]. We treat a variational formulation of the equations of linear elasticity introduced in [Fal91] for elements of order $k = 1$ up to $k = 3$. We focus here on the case $k = 1$, namely, piecewise linear shape functions. In [Sch99b, Sch99a] its equivalence to a mixed formulation using $\mathcal{P}_2 - \mathcal{P}_0$ finite elements is shown. It is known, that this pair of spaces is stable for Stokes equations ([GR86]). Moreover, stable pairs of finite elements spaces for the Stokes equations are usually

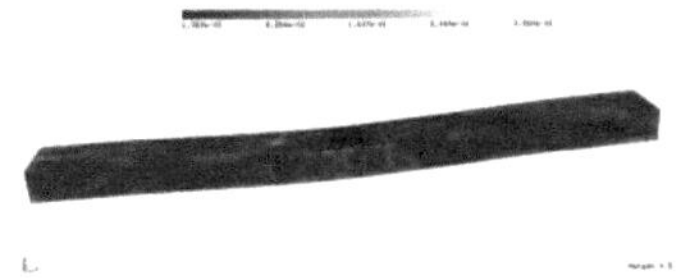

(a) order 1

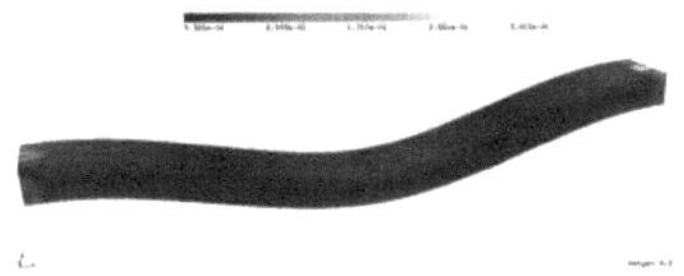

(b) order 2

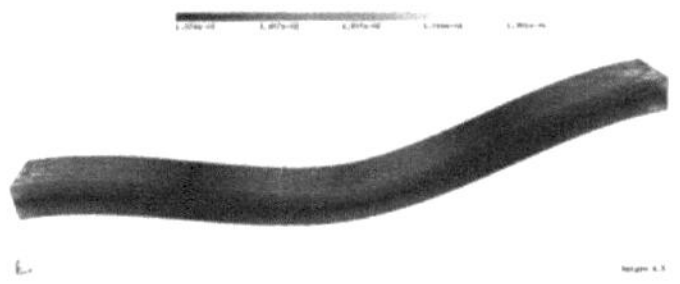

(c) order 3

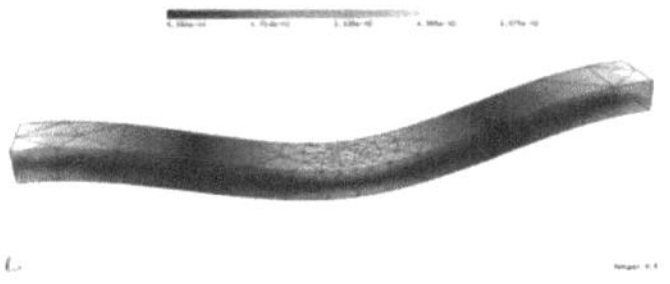

(d) order 4

Figure 2.1: Beam for $\nu = 0.4999$.

locking-free when they are applied to an appropriate mixed formulation of linear elasticity problems (see [GR86, BF91]).

Let us consider a quasi-uniform triangulation $\mathcal{T}_H$ of the domain Ω into d-simplices. For convenience we describe the method for the case $d = 2$ in which we subdivide each of the elements $T \in \mathcal{T}_H$ into four congruent triangles by adding the midpoints of the edges to the set of vertices. The obtained refined triangulation $\mathcal{T}_h$ of Ω has a meshsize $h = H/2$. We introduce the vector spaces $\boldsymbol{V} = [H^1(\Omega)]^2$ and the discrete subspace $\boldsymbol{V}_h := [V_h^1]^2$ of vector-valued nodal piecewise linear basis functions on the fine mesh $\mathcal{T}_h$. By $\hat{\boldsymbol{V}}_h$ we designate $\boldsymbol{V}_h \cap \hat{\boldsymbol{H}}^1(\Omega)$. Then we derive the discrete variational problem (2.33) from (2.42) with

$$a(\boldsymbol{u}_h, \boldsymbol{v}_h) := 2\mu \left\{ (\varepsilon(\boldsymbol{u}_h), \varepsilon(\boldsymbol{v}_h))_0 + \frac{\nu}{1-2\nu} (P_0 \operatorname{div} \boldsymbol{u}_h, P_0 \operatorname{div} \boldsymbol{v}_h)_0 \right\}, \tag{2.52}$$

and

$$L(\boldsymbol{v}_h) := (\boldsymbol{f}, \boldsymbol{v}_h)_0 + \int_{\partial\Omega} \boldsymbol{t}_N \cdot \boldsymbol{v}_h \, \mathrm{d}s \tag{2.53}$$

where $\boldsymbol{f} \in [L_2(\Omega)]^2$ and $\boldsymbol{t}_N \in [L_2(\partial\Omega)]^2$. Moreover, the solution $\boldsymbol{u}_h$ is sought in the space $\hat{\boldsymbol{V}}_h$. The operator P_0 is the L^2-projection onto S_H, that is,

$$P_0(v)|_{T_H} = \frac{1}{|T_H|} \int_{T_H} v \, \mathrm{d}\boldsymbol{x} \qquad \forall T_H \in \mathcal{T}_H, \tag{2.54}$$

for any scalar function $v \in L^2(\Omega)$. This form goes back to [Fal91]. In [Fal91] it is shown that we obtain optimal order error estimates.

For $d = 3$, in [GR86, Subsection II.2.3] it is shown that piecewise $\mathcal{P}_1$-functions together with face bubbles (subset of piecewise $\mathcal{P}_3$-functions) for the velocities and piecewise constant pressure yield stable discretization schemes for Stokes equations, and hence, a locking-free discretization for linear elasticity problems.

2.4.3 Discontinuous Galerkin formulations

Another possible variational formulation of the equations (2.42) using only the displacement $\boldsymbol{u}$ is given in [HL03]. By means of a discontinuous Galerkin (DG) approximation using a penalty term which penalizes the discontinuities in a special way one obtains a-priori error estimates of optimal order which are uniform in the Poisson ratio ν and hence, locking is avoided. More specifically, the approach is an *interior penalty* approach (IPDG), whereas we are only interested in the symmetric case, the *Symmetric Interior Penalty Galerkin* (SIPG) method yielding positive definite stiffness matrices. For a summary on DG methods we may refer to the book [HW08] or the article [ABCM01], which summarizes the basic IPDG approaches and some more.

We consider problem (2.42) in dimension $d = 2, 3$ with $\boldsymbol{u}_D = \boldsymbol{0}$. The triangulation $\mathcal{T}_h$ is assumed to be regular. The set of faces (faces for $d = 3$ and edges for $d = 2$) $\mathcal{E}_h$ is subdivided into

$$\mathcal{E}_h = \mathcal{E}_h^o \cup \mathcal{E}_h^D \cup \mathcal{E}_h^N \quad \text{with} \quad \mathcal{E}_h^o = \mathcal{E}_h \cap \Omega,\ \mathcal{E}_h^D = \mathcal{E}_h \cap \Gamma_D,\ \mathcal{E}_h^N = \mathcal{E}_h \cap \Gamma_N\,.$$

As the name of the DG method suggests we use, instead of functions in $[H^1(\Omega)]^d$, the space of piecewise H^1-functions, called broken spaces. For more details on the general variational form we refer to [AGKZ11]. We immediately switch to the finite dimensional case and define the space $\boldsymbol{V}_h^{DG}$ of piecewise linear functions

$$\boldsymbol{V}_h^{DG} := \{\boldsymbol{v} \in [L^2(\Omega)]^d \,:\, \boldsymbol{v}|_T \in [\mathcal{P}_1(T)]^d \quad \forall T \in \mathcal{T}_h\}\,. \tag{2.55}$$

As in the definition of $\boldsymbol{V}_{h,0}^{RT}$ we regard for each face E a fixed unit normal vector $\boldsymbol{n}_E$. Let T_E^+ be the element for which $\boldsymbol{n}_E$ is the outward vector at face E, while for the element T_E^- $\boldsymbol{n}_E$ is directed inwards. Similarly, we define the traces of a function $\boldsymbol{v}$ at E with $\boldsymbol{v}^+$ or $\boldsymbol{v}^-$ if the values belong to T_E^+ or to T_E^-. Now, let us define the jump and the average operator for a given function $\boldsymbol{w}$, for which the traces on the single elements T_E^+ and T_E^- at the face E are properly defined (piecewise H^1-functions for instance), as

$$[\![\boldsymbol{w}]\!] := (\boldsymbol{w}^+ - \boldsymbol{w}^-) \quad \text{and} \quad \{\!\{\boldsymbol{w}\}\!\} := \left(\frac{\boldsymbol{w}^+ + \boldsymbol{w}^-}{2}\right). \tag{2.56}$$

On boundary faces $E \in \mathcal{E}_h^D \cup \mathcal{E}_h^N$ we specify $[\![\boldsymbol{w}]\!] := \boldsymbol{w}$ and $\{\!\{\boldsymbol{w}\}\!\} := \boldsymbol{w}$. Following [HL03] we define

$$\begin{aligned} a(\boldsymbol{u}_h, \boldsymbol{v}_h) \quad := \quad & \sum_{T\in\mathcal{T}_h} (C\boldsymbol{\varepsilon}(\boldsymbol{u}_h)\,,\, \boldsymbol{\varepsilon}(\boldsymbol{v}_h))_{0,T} \\ & - \sum_{E\in\mathcal{E}_h^o\cup\mathcal{E}_h^D} \left\{(\{\!\{C\boldsymbol{\varepsilon}(\boldsymbol{u}_h)\cdot\boldsymbol{n}_E\}\!\}\,,\,[\![\boldsymbol{v}_h]\!])_{0,E} + (\{\!\{C\boldsymbol{\varepsilon}(\boldsymbol{v}_h)\cdot\boldsymbol{n}_E\}\!\}\,,\,[\![\boldsymbol{u}_h]\!])_{0,E}\right\} \\ & +(2\mu+\lambda)\gamma_0 \sum_{E\in\mathcal{E}_h^o\cup\mathcal{E}_h^D} (\frac{1}{h}[\![P_0\boldsymbol{u}_h]\!]\,,\,[\![P_0\boldsymbol{v}_h]\!])_{0,E} \\ & +2\mu\gamma_1 \sum_{E\in\mathcal{E}_h^o\cup\mathcal{E}_h^D} (\frac{1}{h}[\![\boldsymbol{u}_h]\!]\,,\,[\![\boldsymbol{v}_h]\!])_{0,E} \end{aligned} \tag{2.57}$$

together with the linear functional

$$L(\boldsymbol{v}_h) := (\boldsymbol{f}\,,\,\boldsymbol{v}_h)_{0,\Omega} + \sum_{E\in\mathcal{E}_h^N} (\boldsymbol{t}_N\,,\,\boldsymbol{v}_h)_{0,E}\,. \tag{2.58}$$

The last two terms of (2.57) penalize the jumps across the interior and Dirichlet faces. The operator P_0 is the L^2-projection onto constants along an edge $E \in \mathcal{E}_h$, that is

$$P_0(v)|_E = \frac{1}{|E|}\int_E v\,\mathrm{d}\boldsymbol{x} \qquad \forall E \in \mathcal{E}_h\,. \tag{2.59}$$

The parameters $\gamma_0 > 0$ and $\gamma_1 > 0$ have to be chosen properly to prove stability. In [AGKZ11] it is shown that this bilinear form is consistent in the sense that

$$a(\boldsymbol{u} - \boldsymbol{u}_h, \boldsymbol{v}_h) = 0 \qquad \forall \boldsymbol{v}_h \in \boldsymbol{V}_h^{DG}\,,$$

with $\boldsymbol{u}$ being the exact solution of (2.42) with (2.43). For a-priori error estimates a mesh dependent energy norm is used which is defined by

$$|\!|\!|\boldsymbol{v}|\!|\!|^2 := \sum_{T\in\mathcal{T}_h} (C\boldsymbol{\varepsilon}(\boldsymbol{v})\,,\,\boldsymbol{\varepsilon}(\boldsymbol{v}))_{0,T} + \sum_{E\in\mathcal{E}_h^I\cup\mathcal{E}_h^D} (\frac{1}{h}[\![\boldsymbol{v}]\!]\,,\,[\![\boldsymbol{v}]\!])_{0,E}\,. \tag{2.60}$$

With this energy norm it can be shown, using an elliptic regularity estimate, that the method does not lock due to

$$|||\boldsymbol{u}-\boldsymbol{u}_h||| + ((2\mu+\lambda)(\gamma_0 - c_0))^{1/2}\left[\sum_{E\in\mathcal{E}_h} \|h^{-1/2}\,[\![P_0\boldsymbol{u}_h]\!]\|^2_{0,E}\right]^{1/2}$$
$$\leq Ch(\|\boldsymbol{f}\|_{L^2(\Omega)} + \|\boldsymbol{t}_N\|_{H^{1/2}(\Gamma_N)}), \quad (2.61)$$

from which the L^2-estimate

$$\|\boldsymbol{u}-\boldsymbol{u}_h\|_{L^2(\Omega)} \leq Ch(\|\boldsymbol{f}\|_{L^2(\Omega)} + \|\boldsymbol{t}_N\|_{H^{1/2}(\Gamma_N)}) \quad (2.62)$$

follows. In either of the estimates (2.61) and (2.62) the constant C is independent of the Poisson ratio ν or the Lamé constant λ. For the first estimate we need that $\gamma_0 > c_0$ with c_0 sufficiently large and additionally, $\gamma_1 \geq c_1 > 0$ is necessary.

2.4.4 Mixed formulations

Finally, we consider some mixed formulations of (2.42) in order to give a complete picture of possible variational formulations of the system (2.42). We state 3 standard formulations. When considering mixed formulations of the equations of linear elasticity an additional unknown is introduced. The new unknown can be interpreted as a pressure p or it is given by the stress tensor $\boldsymbol{\sigma}$. The relation between the displacement $\boldsymbol{u}$ and the extra unknown supplies us with another equations. Generally, the elasticity system in mixed form (see [BF91]) looks as:

Find $(u, p) \in V_g \times Q_g$ such that

$$\begin{aligned} a(u, v) &+ b(v, p) &= \langle F_1, v\rangle \qquad &\text{for all } v \in V_0\,, \\ b(u, q) &- c(p, q) &= \langle F_2, q\rangle \qquad &\text{for all } q \in Q_0\,, \end{aligned} \quad (2.63)$$

with V_g, $V_0 \subset V$ and Q_g, $Q_0 \subset Q$ where V and Q are suitable Hilbert spaces.

The first system is obtained if one introduces a new variable $p := \lambda \operatorname{div} \boldsymbol{u}$. Then the bilinear forms $a(.,.)$, $b(.,.)$ and $c(.,.)$ of (2.63) are provided by

$$a(\boldsymbol{u}, \boldsymbol{v}) := 2\mu\,(\varepsilon(\boldsymbol{u})\,,\,\varepsilon(\boldsymbol{v}))_{0,\Omega}\,, \quad b(\boldsymbol{u}, q) := (q\,,\,\operatorname{div}\boldsymbol{u})_{0,\Omega}\,, \quad c(p, q) := \frac{1}{\lambda}\,(p\,,\,q)_{0,\Omega}\,, \quad (2.64)$$

together with the linear functionals

$$\langle F_1, \boldsymbol{v}\rangle := (\boldsymbol{f}\,,\,\boldsymbol{v})_{0,\Omega} + (\boldsymbol{t}_N\,,\,\boldsymbol{v})_{0,\Gamma_N} \quad (2.65)$$

and $F_2 = 0$. The involved spaces are $V_g = [H^1_{u_D,\Gamma_D}(\Omega)]^d$, $V_0 = [H^1_{0,\Gamma_D}(\Omega)]^d$ and $Q_0 = Q_g = L^2(\Omega)$.

Another possibility is to handle the stress $\boldsymbol{\sigma}$ as a separate variable. Thence, we end up with (2.63) where

$$a(\boldsymbol{\sigma}, \boldsymbol{\tau}) := \left(C^{-1}\boldsymbol{\sigma}\,,\,\boldsymbol{\tau}\right)_{0,\Omega}, \quad b(\boldsymbol{\tau}, \boldsymbol{u}) := -\left(\varepsilon(\boldsymbol{u})\,,\,\boldsymbol{\tau}\right)_{0,\Omega} \tag{2.66}$$

and $c(.,.) = 0$. The right-hand sides are given by $F_1 = 0$ and

$$\langle F_2, \boldsymbol{v}\rangle := -\left(\boldsymbol{f}\,,\,\boldsymbol{v}\right)_{0,\Omega} - \left(\boldsymbol{t}_N\,,\,\boldsymbol{v}\right)_{0,\Gamma_N}. \tag{2.67}$$

In this case we have to choose the spaces $V = V_g = V_0 = H_S(\Omega)$, $Q_g = [H^1_{u_D,\Gamma_D}(\Omega)]^d$ and $Q_0 = [H^1_{0,\Gamma_D}(\Omega)]^d$.

When applying integration by parts in $b(.,.)$ we arrive at (2.63) with $a(.,.)$ as in (2.66),

$$b(\boldsymbol{\tau}, \boldsymbol{u}) := \left(\boldsymbol{u}\,,\,\operatorname{div}\boldsymbol{\tau}\right)_{0,\Omega} \tag{2.68}$$

and $c(.,.) = 0$. Moreover, we find the right-hand sides

$$\langle F_1, \boldsymbol{\tau}\rangle := \left(\boldsymbol{\tau}\boldsymbol{n}\,,\,\boldsymbol{u}_D\right)_{0,\Gamma_D} \quad \text{and} \quad \langle F_2, \boldsymbol{v}\rangle := -\left(\boldsymbol{f}\,,\,\boldsymbol{v}\right)_{0,\Omega}. \tag{2.69}$$

The involved spaces have to be chosen as $V = \{\boldsymbol{\tau} \in H_S(\Omega) \,:\, \operatorname{div}\boldsymbol{\tau} \in [L^2(\Omega)]^d\}$ and $Q = Q_0 = Q_g = [L^2(\Omega)]^3$. The solution $\boldsymbol{\sigma}$ is sought in the space $V_g = \{\boldsymbol{\tau} \in V \,:\, \boldsymbol{\tau}\cdot\boldsymbol{n} = \boldsymbol{t}_N \text{ on } \Gamma_N\}$, while V_0 is given by the homogeneous version of V_g. Note that in this case the Dirichlet boundary condition turns out to be the *natural* one, i.e., the one which occurs directly in the right-hand side of the problem, while the Neumann boundary conditions have to be imposed as *essential* boundary conditions.

After homogenization, i.e., if $V_g = V_0$ and $Q_g = Q_0$, we find that due to the theorem of Brezzi [Bre74] (or in [BF91, Theorem 1.1]) there exists a unique solution (u, p) to the saddle point problem (2.63) for $c(.,.) = 0$, if

1. the bilinear forms $a(.,.)$ and $b(.,.)$ are bounded,
2. $a(.,.)$ is coercive on $\ker B = \{q \in Q_0 \,:\, b(v, q) = 0\ \forall v \in V_0\}$,
3. the so-called *inf-sup condition* holds. That is, there exists a $\beta > 0$ such that

$$\inf_{q\in Q_0}\sup_{v\in V_0}\frac{b(q, v)}{\|q\|_Q\,\|v\|_V} \geq \beta\,. \tag{2.70}$$

In the case of $c(.,.) \neq 0$ we have a unique solution (see [BF91, Proposition 1.4]) if $a(.,.)$ and $c(.,.)$ are symmetric, bounded and coercive on V_0 and Q_0, respectively. Moreover, the inf-sup condition has to hold for $b(.,.)$.
When discretizing the problem by conforming finite element spaces V_h and Q_h one has to verify the inf-sup condition (2.70) since it does not follow from the conformity of the spaces.

For the first choice it can be shown ([BF91, Proposition II.4.1]) that the solution is unique and uniform with respect to λ. But, in order to verify (2.70) for the discretized system, linear continuous finite elements are not sufficient for the displacement (cf. [BF91, IV.3]). In [Fal91] it was shown that for piecewise quadratic non-conforming displacements (or higher order) one finds optimal a priori error estimates uniform in λ.

For the other two possibilities we observe that

$$a(\boldsymbol{\tau}, \boldsymbol{\tau}) \geq c \, \|\boldsymbol{\tau}\|_{0,\Omega} \,. \tag{2.71}$$

with $c = \Lambda_{\min}(C^{-1}) = \frac{1-2\nu}{E}$ for isotropic materials (2.47). Thus we do not have coercivity on V uniformly in $\nu \in [0, 1/2)$. Since $\varepsilon(\boldsymbol{u}) \in H_S(\Omega)$ for $\boldsymbol{u} \in [H^1(\Omega)]^d$ it follows from the first equation of (2.63) with the setting (2.66) that $\boldsymbol{\sigma} = C\boldsymbol{\varepsilon}(\boldsymbol{u}) \in H_S(\Omega)$ and thence, setting (2.66) is equivalent to the pure displacement formulation (2.49). This is why we do except volume locking for the second formulation in the case of low order finite elements. Due to the equivalence we immediately obtain stability for ansatz spaces of order greater or equal to 4.

For $b(.,.)$ as in (2.68) the inf-sup stability can be found in [BF91, Bra01]. Since $\ker B$ consists exactly of those functions that are divergence-free the coercivity constant in (2.71) does not depend on ν and hence it can be used to analyze the almost incompressible case. The construction of a compatible pair of finite element spaces is a sophisticated task, especially due to the symmetry condition of the stress field. Stable piecewise polynomial finite elements were developed in [AW02, AC05] for 2D and 3D, respectively. Note that the space used for the displacements is at least of order 2 and it is additionally enriched by some higher order functions, i.e., it is not piecewise linear.

Let us have a closer look at $b(.,.)$ of the last two mixed formulations. If $\boldsymbol{v} \in H_0^1(\Omega)$ and $\operatorname{div} \boldsymbol{\tau} \in H^{-1}(\Omega)$, which is fulfilled for $\boldsymbol{\tau} \in H_S(\Omega)$, both bilinear forms are equal. If one now chooses the displacements to be only in $H(\operatorname{curl}, \Omega)$ we need for the stresses $\boldsymbol{\tau}$ that $\operatorname{div} \operatorname{div} \boldsymbol{\tau} \in H^{-1}(\Omega)$. In [SS07] this setting was introduced and the solvability and uniqueness of the solution was analyzed. Moreover, finite elements have been constructed with optimal order error estimates. These elements do not suffer from locking, even not for lowest order. Lowest order means that one has piecewise linear Nédélec elements (cf. [Néd80, Néd86]) for the displacements and the

related normal components of the normal-stresses are given by linear functions on the faces, i.e., for each face we have d DOFs.
Additionally, in [Sin09] a hybridized version has been introduced and an additive Schwarz block preconditioner was developed for the Schur complement equation.

In [MW11] the authors treat preconditioners for (systems of) PDEs in mixed form, i.e. in indefinite form. Additionally in the report [MW10] a preconditioner for the Stokes equations is posed which shows a uniform convergence behavior in a numerical example. The same preconditioner turns out to be robust for a stable mixed formulation of elasticity using lowest order Tailor-Hood elements.
In [AFW07] a stable mixed formulation is discussed, where the symmetry of σ is only weakly imposed. The authors provide stable mixed finite elements for this formulation.

Chapter 3

Subspace correction methods

The framework of *subspace correction methods* was introduced by J. Xu in [Xu92]. It deals with variational problems on a Hilbert space V. The considered bilinear form naturally introduces the system operator A. The general methods are based on a decomposition of V into a finite number of subspaces that may overlap but do sum up to V. On each subspace an operator $\tilde{A}_i$, represented through a bilinear form, is chosen to approximate A on the subspace. The general method of subspace correction methods solves the residual equation on each subspace by means of $\tilde{A}_i$ approximately. Furthermore, the correction steps may be applied in parallel or successively, leading to the method of *parallel subspace corrections* (MSSC) or *successive subspace corrections* (MSSC). Most of the known iterative solution methods to solve linear variational problems efficiently, can be considered to be subspace correction methods. In this chapter we present different iterative procedures for solving discretizations of PDEs in terms of subspace correction methods.

In Section 3.1 we provide the general framework of subspace correction methods and the related convergence theory. Afterwards, (A)MG is considered in terms of this framework. We present the general MG algorithms, shortly address the convergence investigations and finally provide a convergence estimate for a specific example employing the theory of MSSC. Then, in Section 3.3 attention is drawn to two-level convergence of MG in terms of an estimate on the condition number of the preconditioned system. In Section 3.4 the auxiliary space method based on the fictitious space lemma is addressed. Next, the framework of domain decomposition methods is discussed. Thereby, especially the Schwarz overlapping DD methods align to the framework of subspace corrections. A convergence estimate in terms of the theory of MSSC is stated for a specific example and moreover, it is compared to classical convergence results of Schwarz methods. Finally, in Section 3.6 the convergence properties of MSSC in the case of two overlapping subspaces are derived and numerical results are presented for a specific test case.

3.1 General framework

The general framework of subspace correction aligns to the theory and notation of general Schwarz methods. The first unifying papers are [Xu92] and also corresponding to Schwarz methods [GO95]. The theory of [Xu92] has been further investigated in [XZ02, CXZ08]. In here, we follow the notation of the works by Xu and Zikatanov.

Let us assume we are given a variational problem: Find $u \in V$ such that

$$a(u, v) = f(v) \qquad \forall v \in V\,, \tag{3.1}$$

with $V \subset H$ being a closed subset of the Hilbert space H with the standard inner product $(.\,,\,.)$. Moreover, we assume that $a(.,\,.) \,:\, H \times H \to \mathbb{R}$ is a SPD and continuous bilinear form, which therefore defines an inner product. If f is a continuous linear functional in H^* the problem is well-posed. Since $a(.,\,.)$ defines an inner product it naturally defines a self-adjoint operator $A \,:\, V \to V$, such that

$$(Av\,,\,w) = a(v,\,w) \qquad \forall v,\,w \in V\,. \tag{3.2}$$

Now, let us split V into a, not necessarily direct, sum of closed subspaces $V_i \subset V$, $i = 1,\,\ldots\,,\,J$, with

$$V = \sum_{i=1}^{J} V_i\,. \tag{3.3}$$

With each subspace V_i we associate a bilinear form $a_i(.,\,.)$ that should be an approximation of $a(.,\,.)$ on V_i. Contrary to the setting used in [XZ02] where all occurring bilinear forms only have to satisfy the inf-sup condition on the corresponding spaces, we devise that all $a_i(.,\,.)$ are coercive on V_i. Moreover, in [XZ02] the original bilinear form $a(.,\,.)$ has to fulfill the inf-sup condition on each of the subspaces V_i, which follows automatically from the coercivity on V. If one wants to include mixed problems then we would have to switch to the inf-sup setting. But since we are only dealing with SPD bilinear forms we can simplify those conditions.

In general we consider an iterative procedure for solving (3.1). That is, we provide an initial guess u^0 and the iterative method produces a sequence of u^l $(l = 0, 1, 2, \ldots)$ that should approximate the exact solution u better and better. For each u^{l-1} we obtain u^l by updating with e, $u^l = u^{l-1} + e$, solving approximately the residual equation

$$a(e, v) = f(v) - a(u^{l-1}, v) \qquad \forall v \in V \,, \tag{3.4}$$

which is as difficult to solve as the original equation (3.1). One has to find now an efficient way to compute a good approximation to e.

Subspace correction methods solve (3.4) approximately on each subspace. Those sub-solutions are added up in a certain way in order to obtain the full approximate error. Therefore, we distinguish between a successive approximation of the actual update or a parallel update leading to the *method of successive subspace correction* (MSSC) or to the *method of parallel subspace corrections* (MPSC), respectively.

Algorithm 3.1.1: MSSC($a(.,.)$, $f(.)$, u^0)

for $l = 1, 2, \ldots$ **do**
  $u^l = u^{l-1}$
  for $i = 1$ *to* J **do**
    Let $e_i \in V_i$ solve
      $a_i(e_i, v_i) = f(v_i) - a(u^l, v_i) \quad \forall v_i \in V_i$
    $u^l \leftarrow u^l + e_i$

In order to analyze Algorithm 3.1.1 we introduce the class of linear operators $T_i : V \to V_i$. They are defined through

$$a_i(T_i v, v_i) = a(v, v_i) \qquad \forall v_i \in V_i \,. \tag{3.5}$$

Since we assumed $a_i(.,.)$ to be SPD T_i is well-defined and $\mathcal{R}(T_i) = V_i$. Further, $T_i : V_i \to V_i$ is isomorphic. A closer look at Algorithm 3.1.1 yields for the error propagation

$$u - u^l = E(u - u^{l-1}) = \ldots = E^l(u - u^0) \,, \tag{3.6}$$

with

$$E = (I - T_J)(I - T_{J-1}) \cdots (I - T_1) \,. \tag{3.7}$$

The method of parallel subspace corrections looks as follows.

Algorithm 3.1.2: MPSC($a(.,.)$, $f(.)$, u^0)

---

for $l = 1, 2, \ldots$ **do**
  for $i = 1$ *to* J **do**
    Let $e_i \in V_i$ solve
      $a_i(e_i, v_i) = f(v_i) - a(u^{l-1}, v_i) \quad \forall v_i \in V_i$
  $u^l = u^{l-1} + \sum_{i=1}^J e_i$

Let us define the operator $T : V \to V$ as

$$T = \sum_{i=1}^{J} T_i\,, \tag{3.8}$$

with which it is easy to see that for Algorithm 3.1.2 we have $u - u^l = (I - T)(u - u^{l-1})$. Due to (3.3), i.e., the sum is closed, we can apply the *open mapping theorem* (see Lemma 2.1.8) and obtain the stability of the decomposition

$$\sup_{\|v\|=1} \inf_{v=\sum_i v_i} \sum_{i=1}^{J} \|v_i\|^2 < \infty\,. \tag{3.9}$$

Since the $T_i : V_i \to V_i$ are SPD isomorphisms on V_i, with (3.3) it can be shown that $T : V \to V$ is an isomorphism on V (cf. [XZ02]). Moreover, in [XZ02] it was shown that the norm of T^{-1} is given by an infimum over all decompositions of the sum of norms of the single T_i^{-1}, i.e.,

$$\left(T^{-1}v\,,\, v\right) = \inf_{\sum_i v_i = v} \sum_{i=1}^{J} \left(T_i^{-1}v_i\,,\, v_i\right)\,. \tag{3.10}$$

This yields to a useful result in order estimate the condition number of the operator, namely

$$(\lambda_{\min}(T))^{-1} = \sup_{\|v\|=1} \inf_{\sum v_i = v} \sum_{i=1}^{J} (T_i^{-1}v_i\,,\, v_i)\,.$$

For the analysis of MSSC we also follow [XZ02]. Let us denote by $T_i^* : V_i \to V_i$ the a-adjoint operator of T_i defined by $a(T_i u_i, v_i) = a(u_i, T_i^* v_i)$ for all $u_i, v_i \in V_i$. Additionally, we define the symmetrization of T_i

$$\bar{T}_i := T_i + T_i^* - T_i^* T_i\,. \tag{3.11}$$

In the case of MSSC we find the following useful result in [XZ02].

Theorem 3.1.1. *Suppose that* (3.3) *holds. Assume also that for all* $i = 1, \ldots, J$, $T_i : V_i \to V_i$

are isomorphic and that there exists a constant $\omega \in (0,2)$ *such that*

$$\|T_i v\|_a^2 \le \omega\,(T_i v\,,\,v)_a \qquad \forall v \in V\,. \tag{3.12}$$

Then the following relation holds

$$\|E\|_a^2 = \|(I-T_J)(I-T_{J-1})\cdots(I-T_1)\|_a^2 = \frac{c_0}{1+c_0} \tag{3.13}$$

where

$$c_0 = \sup_{\|v\|=1}\ \inf_{\sum_i v_i = v} \sum_{i=1}^{J} \left(T_i \bar{T}_i^{-1} T_i^* w_i\,,\,w_i\right)_a < \infty \quad \textit{with} \quad w_i = \sum_{j=i}^{J} v_j - T_i^{-1} v_i\,. \tag{3.14}$$

If the solution operators T_i on each subspace are exact solves, that is $a_i(.,.) = a(.,.)$ on V_i and therefore T_i reduces to the idempotent and a-adjoint operator P_i defined by

$$a(P_i v,\,v_i) = a(v,\,v_i) \qquad \forall v_i \in V_i\,. \tag{3.15}$$

Note that P_i restricted to V_i yields the identity. In the following, let us denote by $\Pi_i : V \to V_i$ the orthogonal projection from V to V_i, with respect to $(.,.)$, given by

$$(\Pi_i v\,,\,v_i) = (v\,,\,v_i) \qquad \forall v \in V\,\forall v_i \in V_i\,. \tag{3.16}$$

Using Π_i, the relation (3.15) can be rewritten as

$$a(P_i v,\,\Pi_i w) = a(v,\,\Pi_i w) \qquad \forall w \in V\,. \tag{3.17}$$

The operator P_i can be written as

$$P_i = (\Pi_i A \Pi_i)^{-1}\Pi_i A \qquad i = 1,\ldots,J\,. \tag{3.18}$$

Note that since A is an isomorphism on V and since V_i is a closed subspace of V the operator $A_i := \Pi_i A \Pi_i$ is an isomorphism on V_i.

For the analysis of MSSC with exact solvers for the subproblems we find the following result in [XZ02].

Theorem 3.1.2. *Suppose that* (3.3) *holds. Then the following relation holds*

$$\|E\|_a^2 = \|(I-P_J)(I-P_{J-1})\cdots(I-P_1)\|_a^2 = \frac{c_0}{1+c_0} \tag{3.19}$$

where

$$c_0 = \sup_{\|v\|_a=1} \inf_{\sum_i v_i = v} \sum_{i=1}^{J} \|P_i \sum_{j=i+1}^{J} v_j\|_a^2 < \infty \,. \tag{3.20}$$

Note that the factor ω of (3.12) does not occur in the identity (3.13). This is somehow hidden in the derivation of c_0. However, in the results derived in [Xu92] and [GO95] ω enters the estimates of the norm of the error propagation operator.

3.2 (A)MG viewed as subspace correction

Multigrid is a very powerful tool in order to solve discretizations of PDEs approximately. Their investigation goes back to the 60s by Fedorenko and Bakhvalov. In the 70s Brandt further investigated the method and in the work of Hackbusch, beginning in the late 70s, a lot has been done on the convergence theory of multigrid methods. Since then a lot of people have contributed to this topic. Information about multigrid methods can be found in [TOS01, HT82, Hac85, Bra01, ...].

Generally a multigrid method is a multilevel method, with a nested sequence of subspaces of the whole space V, given by,

$$V_1 \subset V_2 \subset \ldots \subset V_J = V \,, \tag{3.21}$$

which are in the simplest case finite element spaces according to a nested sequence of triangulations $\mathcal{T}_k = \mathcal{T}_{h_k}$. The method aims in the solution $u_h \in V_h$ of

$$a(u_h, v_h) = f(v_h) \qquad \forall v_h \in V_J \,. \tag{3.22}$$

Multigrid is based on the concepts of *smoothing* and *coarse grid correction*. That is, we apply a fast relaxation procedure on the fine grid and on the coarse grid we correct the remaining error by a more advanced method. The basis for the multigrid method to work is an excellent concurrence between those two parts. That is, an error that is not reduced by the relaxation has to be efficiently dealt with by the coarse grid part and vice versa.

Smoothing procedures are for example damped Jacobi or Gauss-Seidel iteration. In a two-level setting (with the spaces $V_H \subset V_h$) the smoother acts on the fine grid approximation by means of the affine linear operator $S : V_h \to V_h$. One might apply several pre- and post-smoothing steps ν_1 and ν_2. In terms of finite element spaces or vector-valued representation of the functions, respectively, we do have a basis transformation $P : V_H \to V_h$, being for nested subspaces simply the embedding from V_H into V_h. P is often called *prolongation*. We also need a *restriction* operator $R : V_h \to V_H$ that we choose to be $R = P^T$, since otherwise the coarse grid operator $A_H = RA_hP$ would be non-symmetric which would not align with the assumptions made so far. The 2-grid algorithm written in terms of operators reads: Its

Algorithm 3.2.1: 2-grid algorithm: $\boldsymbol{y} = MG2(\boldsymbol{b}, \boldsymbol{x})$

$\boldsymbol{x} \leftarrow S^{\nu_1}(\boldsymbol{x})$	// apply ν_1 pre-smoothing steps
$\boldsymbol{d}_h = \boldsymbol{b} - A\boldsymbol{x}$	// compute defect
$\boldsymbol{d}_H = P^T\boldsymbol{d}_h$	// restrict defect to coarse level
$\boldsymbol{v}_H = (P^TAP)^{-1}\boldsymbol{d}_H$	// solve the defect equation on the coarse grid
$\boldsymbol{v}_h = P\boldsymbol{v}_H$	// prolongate the coarse grid update to the fine level
$\boldsymbol{x} \leftarrow \boldsymbol{x} + \boldsymbol{v}_h$	// update the solution
$\boldsymbol{y} = S^{\nu_2}(\boldsymbol{x})$	// apply ν_2 post-smoothing steps

error propagation is given by $\boldsymbol{e}^{(k+1)} = (I - B_{\mathrm{MG},2}^{-1}A)\boldsymbol{e}^{(k)}$. We want to consider only the case $\nu_1 = \nu_2 = 1$ of a single pre- and post-smoothing step. The error propagation is defined via the composition of pre-smoothing, the coarse grid correction and post-smoothing. The error propagation operator of the coarse grid correction step is given by

$$I - PA_H^{-1}P^TA \tag{3.23}$$

and the error propagation operator of the smoothing (relaxation) process is

$$S = I - M^{-1}A\,. \tag{3.24}$$

Here M^{-1} is an approximation to A^{-1}. In order to obtain a symmetric method, the matrices for the pre- and post-smoothing steps are adjoint to each other. Thence, we find $B_{\mathrm{MG},2}^{-1}$ from

$$E_{\mathrm{MG},2} = I - B_{\mathrm{MG},2}^{-1}A = (I - M^{-1}A)(I - PA_H^{-1}P^TA)(I - M^{-T}A)\,. \tag{3.25}$$

Similarly to the computation in [Not05], we find with $C = (I - M^{-1}A)(I - PA_H^{-1}P^TA)$ that

$$\begin{aligned} \|E_{\mathrm{MG},2}\|_A &= \|C(I - M^{-T}A)\|_A = \|CA^{-1}C^TA\|_A \\ &= \|A^{1/2}CA^{-1}C^TA^{1/2}\| = \|(A^{1/2}CA^{-1/2})(A^{1/2}CA^{-1/2})^T\| \\ &= \|(A^{1/2}CA^{-1/2})\|^2 = \|C\|_A^2\,. \end{aligned} \tag{3.26}$$

That is, in order to investigate the error propagation we can neglect pre-smoothing. Written in terms of the subspace correction framework E looks like

$$E = (I - T)(I - P_H) \tag{3.27}$$

with $T_2 = T = M^{-1}A$. $T_1 = P_H$ is the a-orthogonal projection onto V_H, that is, $a_1(.,.) = a(.,.)$. According to Theorem 3.1.1 in [Zik08] it was shown that

$$c_0 = \sup_{v \in V} \frac{\|(I - \bar{\Pi}_H)v\|^2_{\bar{M}}}{\|v\|^2_A} - 1\,, \tag{3.28}$$

where $\bar{\Pi}_H$ is an $(.,.)_{\bar{M}}$-orthogonal projection onto V_H with the symmetrization $\bar{M}$ of M as in 3.11. With $K = c_0 + 1$ one can show that (3.28) is equivalent to the K derived in [FVZ05, Theorem 4.1] which has been proven by direct algebraic computations. Similar derivations have been conducted in [Vas08, Section 6.6]. There, the effect of coarsening by *compatible relaxation* has been investigated. In AMG compatible relaxation means that the coarsening is solely based on the smoother M, and hence, a coarse grid correction which is orthogonal to the (symmetrized) smoother is somehow optimal.

In order to set up a multilevel procedure, let us consider the nested sequence of spaces (3.21). Therefore, we identify the components of level l with subscript (l). That is $A_{(l-1)} := P^T_{(l)} A_{(l)} P_{(l)}$ for $l = 2, \ldots, J$ with $A_{(J)} := A$. By $n_L = J - 1$ we denote the number of coarse levels. Additionally, let η be an integer parameter that determines the (A)MG cycle. The multigrid algorithm, defining the preconditioner B^{-1}_{MG}, is as follows. The choice $\eta = 1$ results in

Algorithm 3.2.2: multigrid algorithm: $\boldsymbol{y} = MG(\boldsymbol{b}, \boldsymbol{x}, l)$

```
if l = 1 then
    x = A_(l)^{-1} b                       // solve exactly on the coarsest level
else
    for s = 1 to η do                      // perform η cycles
        x ← S^{ν1}(x)                      // apply ν1 pre-smoothing steps
        d_(l) = b − A_(l) x                // compute defect
        d_(l−1) = P_(l)^T d_(l)            // restrict defect to coarse level
        v_(l−1) = MG(d_(l−1), 0, l − 1)    // solve coarse defect equ.  recursively
        v_(l) = P_(l) v_(l−1)              // prolongate the coarse grid update
        x ← x + v_(l)                      // update the solution
        y = S^{ν2}(x)                      // apply ν2 post-smoothing steps
```

the so-called *V-cycle* while $\eta = 2$ is typically referred to as *W-cycle*. Moreover, we use the notation $\mathrm{V}(\nu_1, \nu_2)$ for a V-cycle with ν_1 pre-smoothing steps and ν_2 post-smoothing steps. An equivalent notation is used for the W-cycle.

The analysis of multigrid was formerly done by two different approaches. The first one, yielding sharp estimates, is the rigorous or *local Fourier analysis* (LFA). Therefore, quadrilateral uniformly refined meshes are used for the analysis. Another approach, in order to have a general convergence theory is based on two properties of the multigrid procedure, to be specific, the *smoothing property* and the *approximation property*. The approximation property tells us how well the finer space can be approximated by the coarser space. On the other hand, the smoothing property determines the quality of the smoothing process, especially on functions being only slightly affected by the coarse grid correction.

We focus on the qualitative convergence results. Therefore, we need the properties mentioned above. The *approximation property* is that there exists a $c_1 > 0$ such that

$$\|(I - P_{i-1})v_i\|^2 \leq \frac{c_1}{\lambda_i} a(v_i, v_i) \qquad \forall v_i \in V_i\,, \tag{3.29}$$

where $\lambda_i = \rho(A_i)$ being the spectral radius of the system operator A_i on level i. The second one, the smoothing property can be written as

$$\frac{c_2}{\lambda_i}(v_i\,,\, v_i) \leq \left(\bar{M}_i^{-1} v_I\,,\, v_I\right) \qquad \forall v_i \in V_i\,, \tag{3.30}$$

for $c_2 > 0$. There, $\bar{M}_i^{-1}$ denotes the special symmetrization, with respect to the standard inner product, of M_i^{-1}, that is

$$\bar{M}_i^{-1} = M_i^{-T} + M_i^{-1} - M_i^{-T} A_i M_i^{-1} = M_i^{-T}(M_i + M_i^T - A_i)M_i^{-1}\,,$$

which is the smoothing operator when applying the transposed and the usual relaxation consecutively, i.e., $I - \bar{M}_i^{-1}A_i = (I - M_i^{-T}A_i)(I - M_i^{-1}A_i)\,.$

The purely algebraic smoothing property according to [Stü01, p. 434] is given by

$$\|Sv\|_A^2 \leq \|e\|_A^2 - \sigma\|Ae\|_{D^{-1}}^2\,, \tag{3.31}$$

with $\sigma > 0$ and D being the diagonal of the operator A. Viewing (3.30) in terms of matrix operators, we find that (3.31) implies the weaker property (3.30) with $c_2 = \sigma$. Equivalence between (3.30) and (3.31) cannot be shown in general.

In [XZ02] the case of a general number of smoothing steps ν_1 has been considered. In this paper no post-smoothing has been discussed, which can be argued by considerations like (3.26) above. In here, we shortly mention the case $\nu_1 = 1$. On each of the subspaces V_i we apply one smoothing step and on the coarsest level(space), here denoted by V_0, we do exactly invert the operator A_0 and therefore, this step is neglected in the error propagation. For $i = 1, \ldots, J$ we find

$$T_i = (I - S_i)P_{(i)}\,, \tag{3.32}$$

where $P_{(i)}$ might be an arbitrary projection operator onto V_i and the relaxation error propagator S_i is equivalently to (3.24) given by $S_i = I - M_i^{-1}A_i$. Again, M_i is an easily invertible approximation to A_i. In [XZ02] $P_{(i)}$ was chosen to be P_i, that is, the A orthogonal projection onto V_i. With this projection and a special choice for the decomposition of v, $v_i = (P_i - P_{i-1})v$, in [XZ02] it was shown that c_0 can be bounded by

$$c_0 \leq \frac{c_1}{c_2}, \tag{3.33}$$

which leads to the following convergence estimate

$$\|E\|^2 \leq \frac{c_1}{c_1 + c_2}. \tag{3.34}$$

Note that the use of P_i as the projection is a severe restriction, which leads to a uniform convergence result for the V-cycle. Usually, having only the smoothing property and the approximation property on each level, i.e., uniform two-grid convergence on each level, does not imply uniform convergence for $\gamma = 1$, whereas for $\gamma = 2$ the (A)MG method converges uniformly (cf. [TOS01, Theorem 3.2.1] for instance). In [BPWX91] it is shown that the V(1,1)-cycle depends on the number of coarse levels n_L, which is the number of subspaces J if no assumptions on the regularity are used (see also [Vas08, Theorem 5.7]). Using stronger regularity assumptions, that is, if a strong approximation property holds, level-independent convergence can be proven for the V-cycle (see [Vas08, Chapter 5]). Furthermore, the variable V-cycle with a variable number of smoothing steps, usually higher for smaller subspaces, on each level i $\nu_{1,i} = \nu_{2,i}$ shows a level-independent convergence behavior ([Vas08, Theorem 5.24]) under weak regularity assumptions.

3.3 Two-level convergence

In this section we investigate the convergence properties of the two-level MG method, see Algorithm 3.2.1, if it is used as a preconditioner for the conjugate gradient (CG) method, and hence the condition number of $B_{\text{MG},2}^{-1}A$, i.e., $\kappa(B_{\text{MG},2}^{-1}A)$, is the decisive measure for the convergence rate. The derivations in this section are based on the results in References [Not05] and [FV04].

First, let us assume to have a splitting of the n DOFs into fine and coarse DOFs, or we split the DOFs corresponding to the coarse mesh and the set of fine nodes is given by its complement. Alternatively, one may think of splitting the function space V into two disjoint subspaces. Hence, we can rearrange the matrix A, or set up the matrix A with respect to the space-splitting, to come up with the following block structure

$$A = \begin{bmatrix} A_{\mathrm{ff}} & A_{\mathrm{fc}} \\ A_{\mathrm{cf}} & A_{\mathrm{cc}} \end{bmatrix}. \tag{3.35}$$

Usually, the interpolation from coarse space to whole set of DOFs has the form

$$P = \begin{bmatrix} P_{fc} \\ I_c \end{bmatrix} \tag{3.36}$$

where I_c is the identity corresponding to the coarse DOFs and P_{fc} is a mapping from the coarse to the fine DOFs. Using this interpolation operator P, we compute the coarse grid matrix

$$\hat{A}_c := A_H = P^T A P = A_{\mathrm{cc}} + A_{\mathrm{cf}} P_{\mathrm{fc}} + P_{\mathrm{fc}}^T A_{\mathrm{fc}} + P_{\mathrm{fc}}^T A_{\mathrm{ff}} P_{\mathrm{fc}} \,. \tag{3.37}$$

The error propagation matrix $I - B_{\mathrm{MG},2}^{-1} A$ is given by (3.25) from which we gain $B_{\mathrm{MG},2}^{-1}$. Using the interpolation (3.36), we set up a basis transformation matrix

$$J = \begin{bmatrix} I_f & P_{\mathrm{fc}} \\ & I_c \end{bmatrix}. \tag{3.38}$$

In view of (3.35) the system matrix A in a hierarchical basis reads

$$\hat{A} := J^T A J = \begin{bmatrix} A_{\mathrm{ff}} & A_{\mathrm{fc}} + A_{\mathrm{ff}} P_{\mathrm{fc}} \\ A_{\mathrm{cf}} + P_{\mathrm{fc}}^T A_{\mathrm{ff}} & \hat{A}_c \end{bmatrix}. \tag{3.39}$$

Note that this basis transformation leaves the ff-block unchanged, while the *cc*-block is equal to the Galerkin coarse grid matrix $\hat{A}_c$.

Now, let us introduce the measure μ, defined by

$$\mu := \max_{z \neq 0} \frac{(z_{\mathrm{f}} - P_{\mathrm{fc}} z_{\mathrm{c}})^T X_{\mathrm{ff}} (z_{\mathrm{f}} - P_{\mathrm{fc}} z_{\mathrm{c}})}{z^T A z}, \tag{3.40}$$

where X_{ff} is the ff-block corresponding to the fine-coarse partitioned matrix

$$X := M(M + M^T - A)^{-1} M^T, \tag{3.41}$$

see [Not05], which is also called the symmetrized smoother. Note that the quantity μ is closely related to the choice of the interpolation matrix P_{fc} of AMGm (see Chapter 4) according to (4.30). Moreover, the smoothing process enters (3.40) via the matrix X_{ff}. Since $A^{-1} - X^{-1} =$

$(I - M^{-T}A)A^{-1}(I - AM^{-1})$ is SPSD it follows that $X - A \geq 0$ in a positive semidefinite sense. Reformulating (3.40) then yields

$$\mu = \max_{(\boldsymbol{d}_\mathrm{f}^T, \boldsymbol{z}_\mathrm{c}^T) \neq \boldsymbol{0}} \frac{\boldsymbol{d}_\mathrm{f}^T X_\mathrm{ff} \boldsymbol{d}_\mathrm{f}}{\begin{bmatrix} \boldsymbol{d}_\mathrm{f} \\ \boldsymbol{z}_\mathrm{c} \end{bmatrix}^T \hat{A} \begin{bmatrix} \boldsymbol{d}_\mathrm{f} \\ \boldsymbol{z}_\mathrm{c} \end{bmatrix}} \geq \max_{(\boldsymbol{d}_\mathrm{f}^T, \boldsymbol{z}_\mathrm{c}^T) \neq \boldsymbol{0}} \frac{\boldsymbol{d}_\mathrm{f}^T A_\mathrm{ff} \boldsymbol{d}_\mathrm{f}}{\begin{bmatrix} \boldsymbol{d}_\mathrm{f} \\ \boldsymbol{z}_\mathrm{c} \end{bmatrix}^T \hat{A} \begin{bmatrix} \boldsymbol{d}_\mathrm{f} \\ \boldsymbol{z}_\mathrm{c} \end{bmatrix}} \geq 1, \tag{3.42}$$

where $\boldsymbol{d}_\mathrm{f} = \boldsymbol{z}_\mathrm{f} - P_\mathrm{fc}\boldsymbol{z}_\mathrm{c}$ is the defect of the interpolation. It is usually assumed that M fulfills the smoothing property (see (3.30) or (3.31)) which implies that X is SPD (see [FV04] or more generally [TOS01]). Standard computations show that for instance, Gauss-Seidel relaxation has this property.

After this short summary, we recall the following theorem, cf. [Not05].

Theorem 3.3.1 (Analysis of MG,2). *Let A be an SPD matrix partitioned into 2×2 block form (see (3.35)) and let P_{fc} be some interpolation matrix. Let $B_{MG,2}$ be the MG preconditioner defined by (3.25), with non-singular and symmetric smoother M such that $\|I - M^{-1}A\|_A < 1$. Let $\hat{A}$ be the matrix defined by (3.39), and let $\hat{\gamma}$ be the CBS constant associated with $\hat{A}$. Let X and μ be defined by (3.41) and (3.40) respectively. Then, we have*

$$\begin{aligned} \kappa(B_{MG,2}^{-1}A) &\leq \mu \\ &\leq \frac{1}{(1-\hat{\gamma}^2)\lambda_{min}(X_{ff}^{-1}A_{ff})} \\ &\leq \frac{1}{(1-\hat{\gamma}^2)(2-\lambda_{max}(M^{-1}A))\lambda_{min}(M_{ff}^{-1}A_{ff})} \end{aligned}$$

Note that Theorem 3.3.1 is the restriction of Theorem 12 in [Not05] to symmetric smoothers. Additionally we have omitted the lower bounds on the condition number, since those expressions are quite involved. For instance, Jacobi relaxation is a symmetric smoother as well as the Gauss-Seidel method is if it is applied in a symmetric way, i.e., consecutive forward and backward smoothing.[1]

The requirement $\|I - M^{-1}A\|_A < 1$ is equivalent to the condition that the error propagation operator S of the smoother M has to define a convergent iterative process. The CBS constant $\hat{\gamma}$ determines the abstract angle between the subspaces corresponding to the 2×2 partitioning of $\hat{A}$. Mind that therein also the quality of interpolation is taken into account.

[1]The smoothing property holds for any composed relaxation process for which each component fulfills this property.

Remark. The best possible–in view of convergence–choice $P_{\mathrm{fc}} = -A_{\mathrm{ff}}^{-1}A_{\mathrm{fc}}$ results in a block-diagonal matrix $\hat{A}$ and therefore $\hat{\gamma} = 0$. However, this choice is computationally far too expensive in most practical applications.

Since in general it is difficult to determine μ, Theorem 3.3.1 provides further condition number bounds that involve the quality of interpolation via $\hat{\gamma}$ and spectral relations between the (sub)matrices (of) A, M and X.

3.4 Auxiliary space preconditioning viewed as subspace correction

The *auxiliary space method* as used here, was introduced by J. Xu in [Xu96]. This method is related to the *fictitious space method* (FSM) and the *fictitious domain method* (see [Nep07, Nep91, Nep92]). The FSM was introduced in [Nep91, Nep92]. The main tool in the analysis of these methods is the *fictitious space lemma* (cf. [Nep92, GO95, HX07]). In our presentation we follow the notation of [HX07] and we consider problem (3.1) on the Hilbert space V. The SPD bilinear form $a(.,.)$ defines an inner product on V. Additionally, let us assume to have the following building blocks:

1. a *fictitious space* $\bar{V}$ being a Hilbert space with energy inner product $\bar{a}(.,.)$,
2. a continuous and surjective *transfer operator* $\Pi : \bar{V} \to V$.

Now, with the operator representations , i.e., the isomorphisms, $A : V \to V'$ and $\bar{A} : \bar{V} \to \bar{V}'$ associated with $a(.,.)$ and $\bar{a}(.,.)$, respectively, we define the fictitious (auxiliary) space preconditioner by

$$B := \Pi \circ \bar{A}^{-1} \circ \Pi^t : V' \to V . \tag{3.43}$$

The operator B is related to a symmetric bilinear form $b(.,.)$. With the assumptions made so far in [HX07] it was easily shown that B is an isomorphism.

Theorem 3.4.1 (Fictitious space lemma). *Assume that Π is surjective and that the following conditions are fulfilled:*

1. $\exists c_0 > 0 \, \forall v \in V \, \exists \bar{v} \in \bar{V}$: $\quad \Pi\bar{v} = v$ *and* $\|\bar{v}\|_{\bar{A}} \leq c_0 \|v\|_A$.
2. $\exists c_1 > 0 \, \forall \bar{v} \in \bar{V}$: $\quad \|\Pi\bar{v}\|_A \leq c_1 \|\bar{v}\|_{\bar{A}}$.

Then

$$c_0^{-2}\|v\|_A \leq a\,(BAv\,,\,v) \leq c_1^2\|v\|_A \qquad \forall v \in V\,. \tag{3.44}$$

The proof of the theorem in this form can be found in [HX07]. This immediately implies that the spectral condition number is bounded by

$$\kappa(BA) = \frac{\lambda_{\max}(BA)}{\lambda_{\min}(BA)} \leq (c_0 c_1)^2\,. \tag{3.45}$$

So basically the idea of the auxiliary space method is to replace the problem (3.1) in the space V by a similar problem on another space $\bar{V}$. As it is always the case in the framework of preconditioning, this similar problem, determined by $\bar{a}(.\,,.)$, should be much easier to solve than the original problem and on the other hand it should approximate problem (3.1) as good as possible. Note that the auxiliary space method allows for much more freedom than conventional preconditioning, since one could approximate $a(.,\,.)$ in a completely different space $\bar{V}$ by $\bar{a}(.,\,.)$.

In the article [HX07] the authors were aiming at the solution of $H(\mathrm{div})$- and $H(\mathrm{curl})$-equations. Therefore, their auxiliary space was a product space.

If one is able to split the space into subspaces that are related to known problems Theorem 3.4.1 may be used as the general theoretical framework to prove convergence. This results in the necessary condition of a stable splitting.

Generally spoken, $\bar{V}$ may be given by

$$\bar{V} = W_1 \times W_2 \times \ldots \times W_J\,, \tag{3.46}$$

with the Hilbert spaces W_j and the inner product $\bar{a}_j(.,\,.)$. Then, $\bar{a}(.,\,.)$ is given by

$$\bar{a}(\bar{v},\,\bar{v}) := \sum_{j=1}^{J} \bar{a}_j(w_j,\,w_j) \qquad \text{where} \quad \bar{v} = (w_1,\,\ldots,\,w_J)\,, \tag{3.47}$$

and the transfer operator Π consists of the single transfer operators $\Pi_j : W_j \to V$, namely

$$\Pi\bar{v} = \sum_{j=1}^{J} \Pi_j w_j\,, \tag{3.48}$$

and hence, with the operators $\bar{A}_j : W_j \to W_j'$ associated with $\bar{a}(.,.)$ the preconditioner B is given by

$$B = \sum_{j=1}^{J} \Pi_j \circ \bar{A}_j^{-1} \circ \Pi_J^t \,. \tag{3.49}$$

In [HX07] it is pointed out that the special choice $W_1 = V$ and $\bar{A}_j$ being a kind of smoother is the special feature of the specific auxiliary space method. However, this is only a special case of the general case which has been described above. Clearly, from Theorem 3.4.1 we find that the following conditions have to be fulfilled.

1. There have to exist bounds $c_j > 0$ that bound the norms of the single transfer operators Π_j, that is
$$\|\Pi_j w_j\|_A \leq c_j \|w_j\|_{\bar{A}_j} \qquad \forall w_j \in W_j \,. \tag{3.50}$$
2. Verify that there is a $c_0 > 0$, such that for every $v \in V$ there exists a $\bar{v} \in \bar{V}$ with $v = \Pi\bar{v}$ and
$$\|\bar{v}\|_{\bar{A}} \leq c_0 \|v\|_A \,. \tag{3.51}$$

Thence, Theorem 3.4.1 implies that

$$\kappa(BA) \leq c_0^2(c_1^2 + \ldots + c_J^2) \,.$$

Obviously, the preconditioner (3.49) is an additive (parallel) subspace correction method, as can be seen from the update in each step, given by

$$u \longleftarrow u + B(f - Au) = u + \sum_{j=1}^{J} \Pi_j \bar{A}_j^{-1} \Pi_j^t (f - Au) \,,$$

whereas, when applying successively for each $j = 1, \ldots, J$

$$u \longleftarrow u + \Pi_j \bar{A}_j^{-1} \Pi_j^t (f - Au) \,,$$

we arrive at the multiplicative version of auxiliary space method using the auxiliary space $\bar{V}$ given by (3.46).

3.5 Domain decomposition

The set of *domain decomposition* (DD) methods is a quite general framework for solving discretizations of elliptic PDEs. DD methods are considered for instance in [TW05, DW90,

Pec08, Pec12, Vas08]. A very good general characterization is given in [TW05, Preface, p. V]:

> *"Domain decomposition generally refers to the splitting of a partial differential equation, or an approximation thereof, into coupled problems on smaller subdomains forming a partition of the original domain."*

In general we split the domain Ω into J open subdomains $\{\Omega_i\}_{i=1,\dots,J}$, such that $\overline{\Omega} = \bigcup_{i=1}^{J} \overline{\Omega_i}$. Now, there exist a lot of different variants of DD methods. The most important characterization is, if the method is an *overlapping* or a *non-overlapping* method. That is, we distinguish between the case where $\{\Omega_i\}_{i=1,\dots,J}$ defines a partition of Ω or if there exist $i \neq j$ such that $\Omega_i \cap \Omega_j \neq \emptyset$. The latter scheme is also referred to *substructuring* DD methods, for which Neumann-Neumann, finite element tearing and interconnecting (FETI), FETI-DP and balancing domain decomposition by constraints (BDDC) methods are the most prominent examples (see [TW05, Pec08, Pec12] and the references therein). More classical approaches are overlapping methods aligning to the framework of alternating Schwarz methods. Let us focus on those methods in the following.

The analysis of Schwarz overlapping DD methods follows the abstract theory of Schwarz methods (cf. [TW05, XZ02]). Therefore, associated with the splitting $\{\Omega_i\}_{i=1}^{J}$ we are given spaces V_i, which might be subspaces of V, and interpolation operators $R_i^t : V_i \to V$ being the adjoints of restriction operators R_i. Additionally, as in Section 3.1, we do assume SPD bilinear forms $a_i(.,.)$ on V_i, which approximate $a(.,.)$ on $R_i^t V_i$.

We aim at solving problem 2.24 by means of Algorithm 3.1.1 or Algorithm 3.1.2. In the abstract theory an important condition on the setting is that $\{V_i\}_{i=1,\dots,J}$ admits a *stable decomposition*. That is, there exists a constant c_s, such that for every $v \in V$ there exist $\{v_i \in V_i\}_{i=1,\dots,J}$ such that $v = \sum_{i=1}^{J} R_i^t v_i$ and

$$\sum_{i=1}^{J} a_i(v_i, v_i) \leq c_s^2 a(v, v) . \tag{3.52}$$

In the case of exact subsolves, i.e., $V_i \subset V$ and $a_i(.,.) = a(.,.)$ we are in the setting of Lemma 2.1.8, i.e., (3.52) is fulfilled due to Lemma 2.1.8. Nevertheless, the main difficulty lies in finding a general splitting such that the constant c_0 is independent of the number of DOFs n, i.e., the mesh size h and the number of subdomains J or, equivalently, the sizes H_i of the subdomains Ω_i.

Now, let us focus on a quasi-uniform triangulation $\mathcal{T}_h$ of the domain Ω into simplexes and let us consider the scalar problem (2.36) with $\Gamma_D = \partial\Omega$, $\boldsymbol{g}_D = \mathbf{0}$. Additionally, let us focus on the specific splitting $\Omega_i := \bigcup_{T \in \mathcal{N}_i^{\mathcal{T}_h}} T$ for each vertex $v_i \in \mathcal{V}_h$.

For illustration purposes only, we choose $V = V_h^1$, and $V_i = \text{span}\{\varphi_i\}$ with φ_i being the basis function corresponding to vertex v_i. Clearly, $a(\varphi_i, \varphi_i) \approx |\varphi_i|_1^2 \approx h^{-2}\|\varphi_i\|_0^2$ holds, where we have omitted the constants, especially on the bounds c_1 and c_2 of (2.37). Now we consider a smooth function $v \in V$ for which we find $a(v, v) \approx \|v\|_0^2$. Hence, the constant c_0^2 is of the order h^{-2}. The previous considerations show that one has to add a global space V_0, called the *coarse space*, to take care of such cases. This means, that the stability condition is extended by V_0.

Another important ingredient for the analysis of overlapping DD methods is the *partition of unity* $\{\theta_i\}_{i=1,\ldots,J}$ defined through

$$0 \leq \theta_i(\boldsymbol{x}) \leq 1 \qquad \forall \boldsymbol{x} \in \Omega \tag{3.53a}$$

$$\text{supp}(\theta_i) \subset \overline{\Omega_i}\,, \tag{3.53b}$$

$$\textstyle\sum_{i=1}^J \theta_i(\boldsymbol{x}) = 1 \qquad \forall \boldsymbol{x} \in \Omega \tag{3.53c}$$

$$\sup_{\boldsymbol{x}\in\Omega_i} |\nabla\theta_i(\boldsymbol{x})| \leq \tfrac{c_\theta}{\delta_i}\,, \tag{3.53d}$$

where δ_i denotes the thickness of the overlaps of Ω_i with other Ω_j and c_θ is a constant being independent of δ_i and H_i. Additionally, the *finite covering* property means that there exists $N_c \in \mathbb{N}$ such that at most N_c subdomains overlap each other.

Let us come back to the setting discussed above. There, $N_c = 3$ for $d = 2$ and $N_c = 4$ in the three-dimensional case. Now, we choose the spaces $V = H_0^1(\Omega)$, $V_0 = V_h^1$ and the subspaces V_i for $i = 1, \ldots, J$ are chosen as

$$V_i = \{v \in H_0^1(\Omega) \,:\, v(\boldsymbol{x}) = 0 \text{ for } \boldsymbol{x} \in \Omega\backslash\Omega_i\}\,.$$

This example is discussed in [XZ02]. As partition of unity we choose the basis functions of V_0, i.e., $\theta_i = \varphi_i$, which fulfills conditions (3.53). Since the overlap is of size h we obtain $\max_{\boldsymbol{x}\in\overline{\Omega_i}} |\nabla\theta_i(\boldsymbol{x})| \leq c_\theta h^{-1}$. A final ingredient for the set up and analysis is the Clément-type L^2-projection $Q_0 \,:\, V \to V_0$ (cf. [BS07, Section 4.8]) satisfying

$$h^{-1}\|v - Q_0 v\|_0 + |v - Q_0 v|_1 \leq c_I |v|_1\,. \tag{3.54}$$

The following decomposition is used in [XZ02]

$$v_0 = Q_0 v_0\,, \qquad v_i = \theta_i(v - Q_0 v)\,, \qquad \forall i = 1, \ldots, J\,,$$

and additionally, exact solutions of the subproblems are considered. The MSSC is analyzed, for which the authors show that c_0 defined by (3.20) satisfies the bound

$$c_0 \leq \frac{c_2}{c_1}\tilde{c}\,, \tag{3.55}$$

where c_1 and c_2 (see (2.37)) are the bounds of the diffusion parameter. One may compute the constant $\tilde{c}$, given by

$$\tilde{c} = c_I \max\{1 + 2N_c,\, 2c_\theta^2 N_c\}\,. \tag{3.56}$$

Now, let us verify condition (3.52) for the considered setting. We find

$$\begin{aligned}
\sum_{i=0}^{J} \|v_i\|_{a,\Omega}^2 &= \|Q_0 v\|_{a,\Omega}^2 + \sum_{i=1}^{J} \|\theta_i (v - Q_0 v)\|_{a,\Omega_i}^2 \leq c_2 \left(|Q_0 v|_{1,\Omega}^2 + \sum_{i=1}^{J} |\theta_i (v - Q_0 v)|_{1,\Omega_i}^2 \right) \\
&\leq c_2 \left(|Q_0 v|_{1,\Omega}^2 + 2\sum_{i=1}^{J} \|(\nabla\theta_i)(v - Q_0 v)\|_{0,\Omega_i}^2 + 2\sum_{i=1}^{J} \|\theta_i \nabla (v - Q_0 v)\|_{0,\Omega_i}^2 \right) \\
&\leq c_2 \left(|Q_0 v|_{1,\Omega}^2 + 2c_\theta^2 h^{-2} \sum_{i=1}^{J} \|v - Q_0 v\|_{0,\Omega_i}^2 + 2\sum_{i=1}^{J} |v - Q_0 v|_{1,\Omega_i}^2 \right) \\
&\leq 2c_2 \left((1 + c_I^2)|v|_{1,\Omega}^2 + c_\theta^2 h^{-2} N_c \|v - Q_0 v\|_{0,\Omega}^2 + N_c |v - Q_0 v|_{1,\Omega}^2 \right) \\
&\leq 2c_2 \left(1 + c_I^2 + c_I^2 N_c \max\{c_\theta^2, 1\} \right) |v|_{1,\Omega}^2 \\
&\leq \frac{2c_2}{c_1} \left(1 + c_I^2 + c_I^2 N_c \max\{c_\theta^2, 1\} \right) \|v\|_{a,\Omega}^2\,,
\end{aligned}$$

that is, $c_s^2 = \frac{2c_2}{c_1}\left(1 + c_I^2 + c_I^2 N_c \max\{c_\theta^2, 1\}\right)$. Now, from Theorem 2.9 in [TW05] we obtain for the error propagation operator E_{DD} of the actual DD method according to Algorithm 3.1.1 that

$$\|E_{DD}\|_a^2 \leq 1 - \frac{1}{(2N_c^2 + 1)c_s^2}\,.$$

When comparing the latter result we find that it is slightly worse due to factor in front of c_s^2, because c_s^2 is of the same order than the bound (3.55) with (3.56) for c_0.

The previous computations where computed for V_i being the restriction of $H_0^1(\Omega)$ to Ω_i. Nevertheless, we want to solve problem (2.36) and hence, we have to discretize $H_0^1(\Omega)$. Due to the analysis every possible discretization ma be employed as long as $V_0 = V_h^1$. That is, the fine discretization has to align with $\mathcal{T}_h$. One possibility for doing so is to use higher order polynomials on $T \in \mathcal{T}_h$. Alternatively, we could refine the mesh.

In [XZ02] it is observed that high oscillations of the diffusion tensor $\Lambda(\boldsymbol{x})$ do not influence the estimate and hence, they do not significantly alter the convergence behavior of the method. Nevertheless, both estimates depend on $\frac{c_2}{c_1}$ which might be very high for problems with highly heterogeneous coefficients. The aim is to get efficient methods for which the convergence is independent of oscillations in the parameter and of the ratio $\frac{c_2}{c_1}$ (see for instance [Pec12, TW05]).

Finally, note that when using DD methods in an additive form they allow for an parallel implementation. When using exact subspace solvers the operator T defined in (3.8) fulfills ([TW05, Theorem 2.7])

$$\kappa(T) \leq c_s^2(N_c + 1)\,.$$

Hence, if the decomposition is stable with respect to h and J the operator T defined through Algorithm 3.1.2 can be used to devise an efficient PCG iteration.

3.6 Convergence of MSSC with two overlapping subspaces

Let us consider the problem (3.1). We investigate the convergence behavior of the MSSC in the case of two overlapping subspaces, i.e., $V = V_I + V_{II}$ with

$$V_I := V_0 \oplus V_1 \tag{3.57a}$$

$$V_{II} := V_1 \oplus V_2\,, \tag{3.57b}$$

where V_0, V_1 and V_2 being non-overlapping closed subspaces such that the above sums are direct sums, that is, $V = V_0 \oplus V_1 \oplus V_2$. Therefore, any $v \in V$ can be uniquely decomposed into $v = \sum_{i=0}^{2} v_i$ with $v_i \in V_i$, $i = 0, 1, 2$. Thereby, v_i is determined by a suitable projection operator $Q_i : V \to V_i$, i.e., $v_i = Q_i v$. Note that in the case of orthogonal subspaces V_i the operator Q_i would be given by the orthogonal projection Π_i.

We get

$$v = \sum_{i=0}^{2} Q_i v \qquad \forall v \in V\,. \tag{3.58}$$

Having this decomposition in mind, we define the space $\bar{V} := V_0 \times V_1 \times V_2$ endowed with the inner product

$$(\bar{v}\,,\,\bar{w})_{\bar{V}} := \sum_{i=0}^{2} (v_i\,,\,w_i)\,, \tag{3.59}$$

for $\bar{v}$, $\bar{w} \in \bar{V}$. Further, let $\bar{Q} : \bar{V} \to V$ be the mapping defined by $\bar{Q}\bar{v} = \sum_{i=0}^{2} v_i$, i.e., $\bar{Q} = (I, I, I)$. This mapping is bijective. Its inverse is given by $Qv := (Q_0 v, Q_1 v, Q_2 v)^T$ for $v \in V$. Now, let us define $\bar{Q}^t : V \to \bar{V}$ by $\bar{Q}^t = (\Pi_0, \Pi_1, \Pi_2)^T$. Indeed, $\bar{Q}^t$ is the adjoint of $\bar{Q}$, because

$$\left(\bar{Q}\bar{v}\,,\,w\right)=\sum_{i=0}^{2}(v_i\,,\,w)=\sum_{i=0}^{2}(v_i\,,\,\Pi_i w)=\left(\bar{v}\,,\,\bar{Q}^t w\right)_{\bar{V}}\,,$$

for arbitrary $\bar{v}\in\bar{V}$ and $w\in V$. The inner product on each subspace V_i is naturally given by $(.\,,\,.)$.

So far, we find that any $v\in V$ corresponds to exactly one $\bar{v}\in\bar{V}$, where $\bar{v}=Qv$. If the operator A is set up in terms of this space splitting we obtain $\bar{A}:\bar{V}\to\bar{V}$, where $\bar{A}$ is the operator representation of the bilinear form $\bar{a}:\bar{V}\times\bar{V}\to\mathbb{R}$ defined by

$$\bar{a}(.,\,.):=a(\bar{Q}.,\,\bar{Q}.)\,. \tag{3.60}$$

$\bar{A}$ can be written as

$$\bar{A}=\begin{pmatrix}A_{00} & A_{01} & A_{02}\\ A_{10} & A_{11} & A_{12}\\ A_{20} & A_{21} & A_{22}\end{pmatrix}, \tag{3.61}$$

where any A_{ij}, $i,\,j=0,\,1,\,2$ is defined by the operator $A_{ij}\,:\,V_j\to V_i$ with

$$(A_{ij}v_j\,,\,v_i)=a(v_i,\,v_j)=(\Pi_i A\Pi_j v_j,\,v_i)\qquad\forall v_i\in V_i\;\forall v_j\in V_j\,, \tag{3.62}$$

i.e., $A_{ij}=\Pi_i A\Pi_j$. Further, we use $(.\,,\,.)_{\bar{a}}:=\bar{a}(.,\,.)$. Note that

$$\begin{aligned}(\bar{A}\bar{u}\,,\,\bar{v})_{\bar{V}} &= \sum_{i,j=0}^{2}(A_{ij}u_j\,,\,v_i)=\sum_{i,j=0}^{2}(\Pi_i A\Pi_j u_j\,,\,v_i)=\sum_{i,j=0}^{2}(A\Pi_j u_j\,,\,\Pi_i v_i)\\ &= \sum_{i,j=0}^{2}(Au_j\,,\,v_i)=(A\bar{Q}\bar{u}\,,\,\bar{Q}\bar{v})=(\bar{Q}^tA\bar{Q}\bar{u}\,,\,\bar{v})_{\bar{V}}\,.\end{aligned}$$

Hence, the operator $\bar{A}$ is determined by $\bar{A}=\bar{Q}^tA\bar{Q}$. Equivalently, $A=Q^t\bar{A}Q$ with the adjoint $Q^t=(\bar{Q}^t)^{-1}$ of the operator Q. The space splittings of $\bar{V}$ corresponding to (3.57) are $\bar{V}_I:=V_0\times V_1$ and $\bar{V}_{II}:=V_1\times V_2$. The embedding operators $\bar{E}_l\,:\,\bar{V}_l\to\bar{V}$ for $l=I,\,II$ are given by

$$\bar{E}_I=\begin{pmatrix}I & 0\\ 0 & I\\ 0 & 0\end{pmatrix}\quad\text{and}\quad\bar{E}_{II}=\begin{pmatrix}0 & 0\\ I & 0\\ 0 & I\end{pmatrix}.$$

Similarly to (3.59) we define for any $\bar{v}_l,\,\bar{w}_l\in\bar{V}_l$ for $l=I,\,II$

$$(\bar{v}_l\,,\,\bar{w}_l)_{\bar{V}_l}:=\left(\bar{E}_l\bar{v}_l\,,\,\bar{E}_l\bar{w}_l\right)_{\bar{V}}\,. \tag{3.63}$$

For all inner products defined so far we also use the corresponding norms denoted by the same subscript. Moreover, let us define the sub-operators $\bar{A}_I$ and $\bar{A}_{II}$ by

$$\bar{A}_I = \begin{pmatrix} A_{00} & A_{01} \\ A_{10} & A_{11} \end{pmatrix} \quad \text{and} \quad \bar{A}_{II} = \begin{pmatrix} A_{11} & A_{12} \\ A_{21} & A_{22} \end{pmatrix} .$$

In the following we devise (3.20) with respect to the operator representation (3.61) of A. We find

Lemma 3.6.1. *Let $J = 2$ and let the subspaces be given by (3.57). Then in the case of exact subsolves c_0, defined by (3.20), is given by*

$$c_0 = \sup_{\|\bar{v}\|_a = 1} \inf_{v_1 = v_{i,I} + v_{i,II}} \|\bar{E}_I \bar{P}_I \bar{E}_{II} \bar{v}_{II}\|_{\bar{a}}^2 , \tag{3.64}$$

with $\bar{P}_I = \bar{A}_I^{-1} \bar{E}_I^T \bar{A}$.

Proof. Since $J = 2$ we do have

$$c_0 = \sup_{\|v\|_a = 1} \inf_{v = v_I + v_{II}} \|P_I v_{II}\|_a^2 . \tag{3.65}$$

Since V and $\bar{V}$ are isomorphic, we can choose a $\bar{v} \in \bar{V}$ instead of $v \in V$, such that $\|\bar{Q}\bar{v}\|_a = \|\bar{v}\|_{\bar{a}} = 1$. Straightforward computations lead to

$$\begin{aligned} A_I^{-1} &= (\Pi_I A \Pi_I)^{-1} = (\Pi_I Q^t \bar{A} Q \Pi_I)^{-1} \\ &= (\Pi_I Q_I^t \bar{A}_I Q_I \Pi_I)^{-1} = (Q_I \Pi_I)^{-1} \bar{A}_I^{-1} (Q_I \Pi_I)^{-t} = \bar{Q}_I \bar{A}_I^{-1} \bar{Q}_I^t . \end{aligned} \tag{3.66}$$

Thereby, the subscript I denotes the restriction to the subspace $\bar{V}_I$ of the corresponding operators. Moreover, we find

$$\begin{aligned} \|P_I v_{II}\|_a^2 &= (A P_I v_{II} , v_{II}) = (\bar{A} Q P_I v_{II} , Q v_{II})_{\bar{V}} \overset{(3.18)}{=} (\bar{A} Q A_I^{-1} \Pi_I A v_{II} , Q v_{II})_{\bar{V}} \\ &\overset{(3.66)}{=} (\bar{A} Q \bar{Q}_I \bar{A}_I^{-1} \bar{Q}_I^t \Pi_I Q^t \bar{A} Q v_{II} , Q v_{II})_{\bar{V}} = (\bar{A} \bar{E}_I \bar{A}_I^{-1} \bar{E}_I^T \bar{A} Q v_{II} , Q v_{II})_{\bar{V}} , \end{aligned}$$

which is due to $Q\bar{Q}_I = (Q_0, Q_1, Q_2)^T (I, I) = E_I$ and $\bar{Q}_I^t \Pi_I = \bar{Q}_I^t = (\Pi_0, \Pi_1)^T$, since for all $v, w \in V$

$$(\Pi_i \Pi_I v , w) = (\Pi_I v , \Pi_i w) = (v , \Pi_i w) = (\Pi_i v , w) , \tag{3.67}$$

for $i = 0, 1$, because $\Pi_i w \in V_I$. One can easily verify that $\bar{P}_I$ is $\bar{A}$-orthogonal and hence we have $\|P_I v_{II}\| = \|\bar{E}_I \bar{P}_I Q v_{II}\|$. Since the overlap is simply given by V_1, the desired result follows. □

In the analysis of c_0, defined by (3.64), we use the properties of Schur complements and the respective CBS constant. Therefore, let us define the Schur complement between V_I ($\bar{V}_I$ respectively) and V_2 in terms of V_2

$$S_{I,2}^{(2)} := A_{22} - (A_{20},\, A_{21})\, \bar{A}_I^{-1} \begin{pmatrix} A_{02} \\ A_{12} \end{pmatrix} . \tag{3.68}$$

Additionally we need the Schur complements $S_{i,j}^{(j)}$ between V_i and V_j, $i, j = 0, 1, 2$ and $i < j$, with respect to V_j, that is

$$S_{i,j}^{(j)} := A_{jj} - A_{ji} A_{ii}^{-1} A_{ij} . \tag{3.69}$$

With $S_{I,2}$ and $S_{i,j}$ the CBS constants $\gamma_{I,2}$ and $\gamma_{i,j}$ can be found through (see (1.4) or [Axe94])

$$1 - \gamma_{I,2}^2 = \inf_{v_2 \in V_2} \frac{(S_{I,2}^{(2)} v_2\,,\, v_2)_{V_2}}{(A_{22} v_2\,,\, v_2)_{V_2}} , \tag{3.70}$$

$$1 - \gamma_{i,j}^2 = \inf_{v_j \in V_j} \frac{(S_{i,j}^{(j)} v_j\,,\, v_j)_{V_j}}{(A_{jj} v_j\,,\, v_j)_{V_j}} . \tag{3.71}$$

Note that in [Axe94] (Section 9.1) the CBS constant is defined in the finite dimensional setting for a block-partition of a matrix A. Through a careful look one finds, that the theory holds also true in our setting, i.e., a SPD operator A on a Hilbert space V or on one of its subspaces V_i, $i = 0,\, 1,\, 2$.

Now, we find that the constant c_0 of Theorem 3.1.1 is given by the following identity.

Theorem 3.6.2. *The constant* c_0 *in* (3.64) *is given by*

$$c_0 = \sup_{v_2 \in V_2} \frac{(S_{1,2}^{(2)} v_2\,,\, v_2)_{V_2}}{(S_{I,2}^{(2)} v_2\,,\, v_2)_{V_2}} - 1 . \tag{3.72}$$

Proof. Straightforward calculation leads to

$$\bar{E}_I \bar{P}_I = \begin{pmatrix} \bar{A}_I^{-1} & \begin{matrix} 0 \\ 0 \end{matrix} \\ \begin{matrix} 0 & 0 \end{matrix} & 0 \end{pmatrix} \cdot \begin{pmatrix} \bar{A}_I & \begin{matrix} A_{02} \\ A_{12} \end{matrix} \\ \begin{matrix} A_{20} & A_{21} \end{matrix} & A_{22} \end{pmatrix} = \begin{pmatrix} \begin{matrix} I & 0 \\ 0 & I \end{matrix} & \bar{A}_I^{-1} \begin{pmatrix} A_{02} \\ A_{12} \end{pmatrix} \\ \begin{matrix} 0 & 0 \end{matrix} & 0 \end{pmatrix} ,$$

and hence, we arrive at

$$\bar{A}\bar{E}_I\bar{P}_I = \begin{pmatrix} \multicolumn{2}{c}{\bar{A}_I} & \begin{matrix} A_{02} \\ A_{12} \end{matrix} \\ A_{20} & A_{21} & A_{22} \end{pmatrix} \cdot \left(\begin{matrix} I & 0 \\ 0 & I \\ 0 & 0 \end{matrix} \quad \begin{matrix} \bar{A}_I^{-1} \begin{pmatrix} A_{02} \\ A_{12} \end{pmatrix} \\ 0 \end{matrix} \right)$$

$$= \begin{pmatrix} A_{00} & A_{01} & A_{02} \\ A_{10} & A_{11} & A_{12} \\ A_{20} & A_{21} & (A_{20}, A_{21})\,\bar{A}_I^{-1} \begin{pmatrix} A_{02} \\ A_{12} \end{pmatrix} \end{pmatrix} .$$

Finally, using that A and $\bar{A}$ are self-adjoint, we find

$$\bar{P}_I^T \bar{E}_I^T \bar{A}\bar{E}_I\bar{P}_I = \begin{pmatrix} I & 0 & 0 \\ 0 & I & 0 \\ \multicolumn{2}{c}{(A_{20}, A_{21})\,\bar{A}_I^{-1}} & 0 \end{pmatrix} \cdot \bar{A}\bar{E}_I\bar{P}_I = \bar{A}\bar{E}_I\bar{P}_I \,.$$

Now, let us define the operator

$$\tilde{A}_{II} := \bar{A}_{II} - \begin{pmatrix} 0 & 0 \\ 0 & S_{I,2}^{(2)} \end{pmatrix} . \tag{3.73}$$

$\tilde{A}_{II}$ is the block of $\bar{A}\bar{E}_I\bar{P}_I$ corresponding to $\bar{V}_{II}$. Then, equation (3.64) reduces to

$$c_0 = \sup_{\|\bar{v}\|_{\bar{a}}=1} \inf_{v_1=v_{i,I}+v_{i,II}} (\tilde{A}_{II}\bar{v}_{II} \,,\, \bar{v}_{II})_{\bar{V}_{II}} \,. \tag{3.74}$$

Next, using the minimization property of the Schur complement, that is, for fixed $v_2 \in V_2$ we have

$$\inf_{v_{1,II}\in V_1} (\bar{A}_{II}\bar{v}_{II} \,,\, \bar{v}_{II})_{\bar{V}_{II}} = (S_{1,2}^{(2)} v_2 \,,\, v_2)_{V_2} \,,$$

we find that for any $v_2 \in V_2$

$$\inf_{v_1=v_{i,I}+v_{i,II}} (\tilde{A}_{II}\bar{v}_{II} \,,\, \bar{v}_{II})_{\bar{V}_{II}} = ((S_{1,2}^{(2)} - S_{I,2}^{(2)})v_2 \,,\, v_2)_{V_2}$$

holds. Now, the constant c_0 can be alternatively written as the infimum of all $\lambda > 0$ satisfying

$$((S_{1,2}^{(2)} - S_{I,2}^{(2)})v_2 \,,\, v_2)_{V_2} \le \lambda(\bar{A}\bar{v} \,,\, \bar{v})_{\bar{V}} \qquad \forall \bar{v} \in \bar{V} \,. \tag{3.75}$$

In terms of a generalized eigenproblem, we have to find the supremal λ, such that there exists a $\bar{v} \in \bar{V}$ with

$$\begin{pmatrix} 0 \\ 0 \\ (S_{1,2}^{(2)} - S_{I,2}^{(2)})v_2 \end{pmatrix} = \lambda \bar{A} \bar{v}\,. \tag{3.76}$$

Clearly, we do have $\dim(\bar{V}_I)$ generalized eigenvalues that are 0, which might be infinitely many. We need $\lambda \neq 0$ satisfying (3.76). As a side equation we get

$$\bar{A}_I \begin{pmatrix} v_0 \\ v_1 \end{pmatrix} + \begin{pmatrix} A_{02} \\ A_{12} \end{pmatrix} v_2 = \begin{pmatrix} 0 \\ 0 \end{pmatrix} \Leftrightarrow \begin{pmatrix} v_0 \\ v_1 \end{pmatrix} = -\bar{A}_I^{-1} \begin{pmatrix} A_{02} \\ A_{12} \end{pmatrix} v_2\,. \tag{3.77}$$

Inserting (3.77) in (3.76) yields the reduced generalized eigenvalue problem: Find (λ, v_2) with maximal λ such that

$$(S_{1,2}^{(2)} - S_{I,2}^{(2)})v_2 = \lambda S_{I,2}^{(2)} v_2\,, \tag{3.78}$$

or, equivalently,

$$S_{1,2}^{(2)} v_2 = (1+\lambda) S_{I,2}^{(2)} v_2\,. \tag{3.79}$$

This finally leads to

$$c_0 = \sup_{v_2 \in V_2} \frac{(S_{1,2}^{(2)} v_2\,,\, v_2)_{V_2}}{(S_{I,2}^{(2)} v_2\,,\, v_2)_{V_2}} - 1\,. \tag{3.80}$$

□

In the following, we show that c_0 is related to the CBS constant $\hat{\gamma}_{0,2}$ of an operator $\hat{A}$, which is obtained by an elimination of the components corresponding to V_1 of the operator $\bar{A}$. Therefore, let us introduce the operator $\hat{A} : V_0 \times V_2 \to V_0 \times V_2$ with

$$\hat{A} := \begin{pmatrix} S_{0,1}^{(0)} & A_{02} - A_{01} A_{11}^{-1} A_{12} \\ A_{20} - A_{21} A_{11}^{-1} A_{10} & S_{1,2}^{(2)} \end{pmatrix}, \tag{3.81}$$

and denote the corresponding CBS constant by $\hat{\gamma}_{0,2}$. With $X_{210} := A_{20} - A_{21} A_{11}^{-1} A_{10}$ and its adjoint X_{012} the Schur complement $\hat{S}^{(2)}$ of $\hat{A}$ is given by $\hat{S}^{(2)} = S_{1,2}^{(2)} - X_{210} (S_{0,1}^{(0)})^{-1} X_{012}$, and hence $\hat{\gamma}_{0,2}$ is defined through (compare with (1.4))

$$1 - \hat{\gamma}_{0,2}^2 = \inf_{v_2 \in V_2} \frac{(\hat{S}^{(2)} v_2\,,\, v_2)_{V_2}}{(S_{1,2}^{(2)} v_2\,,\, v_2)_{V_2}}\,. \tag{3.82}$$

Lemma 3.6.3. *c_0 given by* (3.64) *fulfills the identity*

$$c_0 = \frac{\hat{\gamma}_{0,2}^2}{1-\hat{\gamma}_{0,2}^2}, \tag{3.83}$$

with $\hat{\gamma}_{0,2}$ specified through (3.82).

Proof. First of all, note that $S_{I,2}^{(2)} = \hat{S}^{(2)}$, which follows from the following derivations. We have

$$\bar{A}_I = \begin{pmatrix} A_{00} & A_{01} \\ A_{10} & A_{11} \end{pmatrix} = \begin{pmatrix} I & A_{01}A_{11}^{-1} \\ 0 & I \end{pmatrix} \begin{pmatrix} S_{0,1}^{(0)} & 0 \\ 0 & A_{11} \end{pmatrix} \begin{pmatrix} I & 0 \\ A_{11}^{-1}A_{10} & I \end{pmatrix}.$$

Hence,

$$\bar{A}_I^{-1} = \begin{pmatrix} I & 0 \\ -A_{11}^{-1}A_{10} & I \end{pmatrix} \begin{pmatrix} (S_{0,1}^{(0)})^{-1} & 0 \\ 0 & A_{11}^{-1} \end{pmatrix} \begin{pmatrix} I & -A_{01}A_{11}^{-1} \\ 0 & I \end{pmatrix},$$

and thus, with (3.68) it follows that

$$\begin{aligned} S_{I,2}^{(2)} &= A_{22} - (A_{20}, A_{21})\, \bar{A}_I^{-1} \begin{pmatrix} A_{02} \\ A_{12} \end{pmatrix} \\ &= A_{22} - \left(A_{20} - A_{21}A_{11}^{-1}A_{10},\, A_{21}\right) \begin{pmatrix} (S_{01}^{(0)})^{-1} & 0 \\ 0 & A_{11}^{-1} \end{pmatrix} \begin{pmatrix} A_{02} - A_{01}A_{11}^{-1}A_{12} \\ A_{12} \end{pmatrix} \\ &= \hat{S}^{(2)}. \end{aligned}$$

Now, we use expression (3.72), where c_0 is expressed as a supremum, together with (3.82).

$$\begin{aligned} c_0 &= \sup_{v_2 \in V_2} \frac{(S_{1,2}^{(2)} v_2\,,\, v_2)_{V_2}}{(S_{I,2}^{(2)} v_2\,,\, v_2)_{V_2}} - 1 \\ &= \frac{1}{\inf_{v_2\in V_2} \frac{(\hat{S}^{(2)} v_2\,,\, v_2)_{V_2}}{(S_{1,2}^{(2)} v_2\,,\, v_2)_{V_2}}} - 1 = \frac{1}{1-\hat{\gamma}_{0,2}^2} - 1 = \frac{\hat{\gamma}_{0,2}^2}{1-\hat{\gamma}_{0,2}^2}. \end{aligned}$$

□

Using (3.83), we are able to show

Corollary 3.6.4. *For the setting* (3.57) *the error propagation operator E satisfies*

$$\|E\|_a^2 = \hat{\gamma}_{0,2}^2. \tag{3.84}$$

Proof. Plugging (3.83) into (3.19) yields the desired result. □

Summarizing, we have that the a-norm of the error propagation operator in case of two overlapping subspaces is given by the CBS constant between V_0 and V_2 of the operator A where V_1 is eliminated in A. This coincides with the theory obtained in [KW88]. Since in the case of no overlap we do have a-orthogonal projections $I - P_i$ and therefore, the MSSC coincides with the method of alternating projections (cf. [Deu01]). In this case, (3.84) yields that the norm of the error propagation operator is $\gamma_{0,2}$. This aligns with [KW88], where this convergence result was obtained for the method of alternating projections if two subspaces are considered.
Note that all considerations so far work for the general case of infinite dimensional spaces.

3.6.1 The finite dimensional case

If V is finite dimensional, the operator A can be represented in matrix form. Therefore, without loss of generality, we do have $V := \boldsymbol{V} = \mathbb{R}^n$, $n \in \mathbb{Z}$. The inner product is the standard l_2-inner product. Further, let $V_i = \boldsymbol{V}_i \subset \boldsymbol{V}$ be the closed subspaces with $\dim \boldsymbol{V}_i = n_i$ and a linear independent basis $\{\boldsymbol{v}_j^i\}_{j=1,\dots,n_i}$ for $i = 0, 1, 2$. The dimensions n_i sum up to n. Let us denote by $B_i = (\boldsymbol{v}_1^i, \dots, \boldsymbol{v}_{n_i}^i) \in \mathbb{R}^{n \times n_i}$ the basis transformation matrices from $\tilde{\boldsymbol{V}}_i := \mathbb{R}^{n_i}$ to $\boldsymbol{V}_i$ and moreover, $B = (B_0, B_1, B_2)$ is the global basis transformation. The transformations $B_I = (B_0, B_1)$ and $B_{II} = (B_1, B_2)$ denote the transformations of the subspaces $\boldsymbol{V}_I$ and $\boldsymbol{V}_{II}$, respectively.

Now, let us reconsider the operators Π_i and Q. Every $\boldsymbol{v}_i \in \boldsymbol{V}_i$ may be be represented via $\tilde{\boldsymbol{v}}_i \in \tilde{\boldsymbol{V}}_i$, such that $\boldsymbol{v}_i = B_i \tilde{\boldsymbol{v}}_i$ and hence relation (3.16) transforms to

$$(\Pi_i \boldsymbol{v} \,,\, B_i \tilde{\boldsymbol{v}}_i) = (\boldsymbol{v} \,,\, B_i \tilde{\boldsymbol{v}}_i) \qquad \forall \boldsymbol{v} \in \boldsymbol{V} \, \forall \tilde{\boldsymbol{v}}_i \in \tilde{\boldsymbol{V}}_i \,,$$

which finally leads to $\Pi_i = B_i (B_i^T B_i)^{-1} B_i^T$. In this case the enlarged space $\bar{V}$ is given by $\bar{V} := \bar{\boldsymbol{V}} = \boldsymbol{V}_0 \times \boldsymbol{V}_1 \times \boldsymbol{V}_2 \subset \mathbb{R}^{3n}$. The operator $Q : \boldsymbol{V} \to \bar{\boldsymbol{V}}$ is

$$Q = \begin{pmatrix} B_0 & 0 & 0 \\ 0 & B_1 & 0 \\ 0 & 0 & B_2 \end{pmatrix} \cdot B^{-1} . \tag{3.85}$$

Now, let us introduce the operator $\tilde{A} := B^T A B$ being the operator A defined in terms of the basis B. Additionally, with $\Pi = (\Pi_0, \Pi_1, \Pi_2)$ the operator $\bar{A} \in \mathbb{R}^{3n \times 3n}$ is given by $\bar{A} = \Pi^T A \Pi$.

Since, $\Pi_i B_i = B_i$ we find that

$$\tilde{A} = \begin{pmatrix} B_0^T & 0 & 0 \\ 0 & B_1^T & 0 \\ 0 & 0 & B_2^T \end{pmatrix} \bar{A} \begin{pmatrix} B_0 & 0 & 0 \\ 0 & B_1 & 0 \\ 0 & 0 & B_2 \end{pmatrix} . \tag{3.86}$$

That is, the matrix $\tilde{A}$ represents exactly the matrix $\bar{A}$ used to derive the constant c_0 in Lemma 3.6, but already in the basis $\{\boldsymbol{v}_j^i\}_{i=0,1,2;\,j=1,\ldots,n_i}$. With this setting the error propagation matrix is given by

$$E = (I - C_{II}A)(I - C_I A) = (I - B_{II}(B_{II}^T A B_{II})^{-1} B_{II}^T A)(I - B_I(B_I^T A B_I)^{-1} B_I^T A) . \tag{3.87}$$

Therefore, $\hat{\gamma}_{0,2}$, defined in (3.82), is given by

$$1 - \hat{\gamma}_{0,2}^2 = \lambda_{\min}({S_{1,2}^{(2)}}^{-1} \hat{S}^{(2)}) , \tag{3.88}$$

where we start with $\tilde{A}$ in order to compute the matrices $S_{1,2}^{(2)}$ and $\hat{S}^{(2)}$.

In order to confirm the results of Corollary 3.6.4 let us consider the following example. Let $n, m, k \in \mathbb{N}_0$ and let Ω be the domain $\Omega := (0,\, 2m+k+2) \times (0,\, n+2)$. Further, let us consider the Laplace problem (2.35), i.e., $f = 0$, with $\Lambda = I_{\mathbb{R}^2}$, $\Gamma_D = \partial\Omega$, $\Gamma_N = \emptyset$ and $g_D = 0$ on Γ_D. We subdivide Ω according to Figure 3.1 into triangles with side length 1 and approximate the PDE by piecewise linear elements, that is, $V_h = V_h^1$. We neglect the DOFs on the boundary

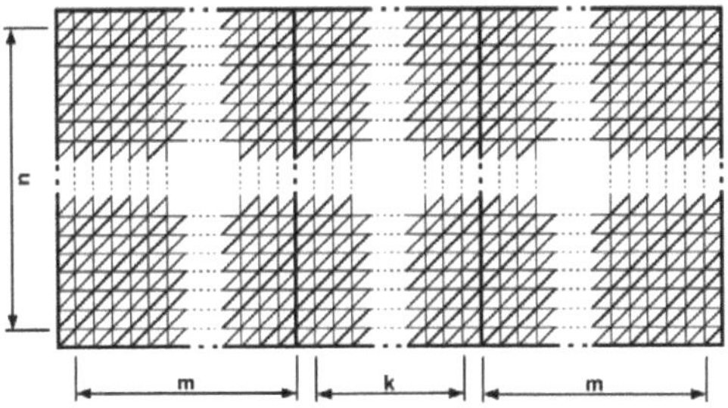

Figure 3.1: Mesh of the general test case for the MSSC.

and numerate the DOFs from top to bottom and left to right. Then, A can be written in block-form as in (3.61) with

$$\begin{aligned} A_{00} = A_{22} &= D_n \otimes I_{\mathbb{R}^m} + I_{\mathbb{R}^n} \otimes D_m \,, \\ A_{11} &= D_n \otimes I_{\mathbb{R}^k} + I_{\mathbb{R}^n} \otimes D_k \,, \\ A_{01} = A_{10}^T &= -E_{m,k}^1 \otimes I_{\mathbb{R}^n} \,, \\ A_{12} = A_{21}^T &= -E_{m,k}^k \otimes I_{\mathbb{R}^n} \,, \\ A_{02} = A_{20}^T &= 0 \,. \end{aligned}$$

Thereby, we use the matrices $D_l \in \mathbb{R}^{l \times l}$ $(l \in \mathbb{N})$, $E_{m,k}^1 \in \mathbb{R}^{m \times k}$ and $E_{m,k}^k \in \mathbb{R}^{m \times k}$ given by

$$\begin{aligned} D_l &:= \begin{pmatrix} 2 & -1 & & & 0 \\ -1 & 2 & -1 & & \\ & \ddots & \ddots & \ddots & \\ & & -1 & 2 & -1 \\ 0 & & & -1 & 2 \end{pmatrix}, \\ E_{m,k}^1 &:= \begin{pmatrix} 1 & 0 & \dots & 0 \\ & 0 & & \end{pmatrix}, \\ E_{m,k}^k &:= \begin{pmatrix} 0 & \dots & 0 & 1 \\ & 0 & & \end{pmatrix}. \end{aligned}$$

In the following tables we compare $\|E\|_a^2$ and the number of iterations of the MSSC for a reduction of the energy-norm of the error by a factor of $\varepsilon = 10^{-10}$ for varying n and m with fixed k. In Table 3.1 no overlap was used, i.e., $k = 0$, while in Table 3.2 and Table 3.3 the overlap was chosen to be $k = 1$ and $k = 4$, respectively. Contrary, in Figure 3.2 the decrease of $\|E\|_a^2$ is depicted for fixed ratios $\frac{n}{m} = 1$, $n = m = 4, 8, 16, 32$, for varying k. Note that the norm of the error propagator for increasing k is exponentially decreasing. From those observations, we see that in order to get a bounded convergence rate, the overlap would have to depend on m and especially on n, which is not a desired property.

However, the numerical experiments reveal that it actually holds

$$\frac{\|E\boldsymbol{u}^k\|_a}{\|\boldsymbol{u}^k\|_a} \leq \|E\|_a^2 \,, \tag{3.89}$$

for $k \geq 1$, which is surprising since one would expect $\|E\|_a$ instead of the bound given above. Finally, we show the following result.

$k = 0$	m											
n	1		2		4		8		16		32	
1	0.06	9	0.07	9	0.07	9	0.07	9	0.07	9	0.07	9
2	0.11	11	0.14	11	0.15	12	0.15	12	0.15	12	0.15	12
4	0.18	14	0.26	18	0.29	19	0.30	19	0.30	18	0.30	18
8	0.22	16	0.37	23	0.47	29	0.50	31	0.50	32	0.50	30
16	0.24	17	0.42	26	0.58	40	0.67	53	0.69	56	0.69	55
32	0.25	17	0.44	27	0.62	46	0.75	73	0.81	98	0.83	105

Table 3.1: $\|E\|_a^2$ and iteration number for $\varepsilon = 10^{-10}$.

$k = 1$	m											
n	1		2		4		8		16		32	
1	0.00	5	0.01	5	0.01	5	0.01	5	0.01	5	0.01	5
2	0.02	6	0.02	6	0.02	7	0.02	7	0.02	6	0.02	6
4	0.05	8	0.07	9	0.09	10	0.09	10	0.09	10	0.09	10
8	0.08	10	0.16	13	0.23	16	0.25	16	0.25	16	0.25	16
16	0.10	11	0.22	15	0.36	22	0.45	27	0.48	29	0.48	28
32	0.11	11	0.24	16	0.42	25	0.58	39	0.66	50	0.68	52

Table 3.2: $\|E\|_a^2$ and iteration number for $\varepsilon = 10^{-10}$.

$k = 4$	m											
n	1		2		4		8		16		32	
1	0.00	3	0.00	3	0.00	3	0.00	3	0.00	3	0.00	3
2	0.00	3	0.00	3	0.00	3	0.00	3	0.00	3	0.00	3
4	0.00	4	0.00	5	0.00	4	0.00	4	0.00	4	0.00	4
8	0.01	6	0.02	6	0.03	7	0.03	7	0.03	7	0.03	7
16	0.02	6	0.05	8	0.10	10	0.14	12	0.16	12	0.16	12
32	0.03	7	0.08	9	0.16	13	0.28	17	0.36	21	0.39	22

Table 3.3: $\|E\|_a^2$ and iteration number for $\varepsilon = 10^{-10}$.

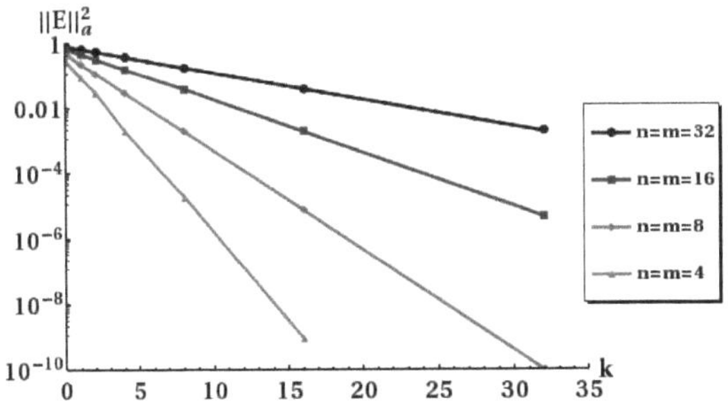

Figure 3.2: $\|E\|_a^2$ for varying k and fixed ratio of n and m, that is $n = m$.

Lemma 3.6.5. *For E defined in* (3.87) *the following holds*

$$\lambda_{\max}(E) = \|E\|_a^2 . \tag{3.90}$$

Proof. Standard calculations yield that $\|E\|_a^2 = \lambda_{\max}(A^{-1}E^T AE)$ and further, using the matrices $P_I = C_I A$ and $P_{II} = C_{II}A$ defined in (3.87), we find

$$\begin{aligned} A^{-1}E^T AE &= A^{-1}(I - P_I^T)(I - P_{II}^T)A(I - P_{II})(I - P_I) \\ &= A^{-1}(I - AC_I)(I - AC_{II})A(I - C_{II}A)(I - AC_I) \\ &= A^{-1}A(I - P_I)(I - P_{II})(I - P_{II})(I - P_I) = (I - P_I)E . \end{aligned} \tag{3.91}$$

Straightforward calculations lead to

$$(I - P_I) = \begin{pmatrix} 0 \ 0 & -\bar{A}_I^{-1}\begin{pmatrix} A_{02} \\ A_{12} \end{pmatrix} \\ 0 \ 0 & \\ 0 \ 0 & I \end{pmatrix},$$

$$(I - P_{II}) = \begin{pmatrix} I & 0 \ 0 \\ -\bar{A}_{II}^{-1}\begin{pmatrix} A_{10} \\ A_{20} \end{pmatrix} & \begin{matrix} 0 \ 0 \\ 0 \ 0 \end{matrix} \end{pmatrix}.$$

Summarizing the eigenvalue problems we find that for $\lambda \neq 0$ the condition $\boldsymbol{v}_2 \neq \boldsymbol{0}$ holds. There are $n_I = n_0 + n_1$ zero eigenvalues and n_2 nonzero ones. The (sub)eigenvalue problem for all $\lambda \neq 0$ is

$$\left[\bar{A}_{II}^{-1}\begin{pmatrix} A_{10} \\ A_{20} \end{pmatrix}\right]_2 \left[-\bar{A}_I^{-1}\begin{pmatrix} A_{02} \\ A_{12} \end{pmatrix}\right]_0 \boldsymbol{v}_2 = \lambda \boldsymbol{v}_2 ,$$

which is the same for E and $(I - P_I)E$. The corresponding components v_0 and v_1 might differ, but the eigenvalues of both operators are the same and therefore, the largest eigenvalue of both matrices coincide. □

Chapter 4

AMG for linear elasticity (AMGm)

We are concerned with the solution of large-scale systems of linear equations

$$Ax = \boldsymbol{b} \tag{4.1}$$

arising from first-order finite element discretization of linear elasticity problems (see Subsection 2.4.1). In this case the system matrix A is symmetric and positive definite (SPD). It is well known that algebraic multigrid (AMG) methods can serve as efficient linear solvers or preconditioners for this type of problems. In particular, AMG using element interpolation (AMGe), see, e.g., [BCF+01, HV01, JV01], AMG based on smoothed aggregation [VMB96, VBM01], and generalized aggregation multilevel solvers [FB97] provide powerful solution tools. Adaptive algebraic multigrid methods [BFM+06] use an adaptive process in which the evolving solver improves its own components. In this way they remove the need to make any specific assumptions on the near null space of the matrix A. The generalized global basis method provides a technique to accelerate a multigrid scheme by an additional coarse grid correction that filters out slowly converging modes [WFTS04, WFTS05]. While these methods in general increase the robustness of classical (algebraic) multigrid algorithms, they also increase, sometimes even substantially, the set-up costs. Recently, in [ZSTB10] an AMG method has been developed for elasticity problems. Thereby, the inherent structure of the elasticity problem is used to set up the AMG components.

A variant of AMGe, called algebraic multigrid based on computational molecules (AMGm), has been proposed in [Kra08] (see also its predecessor [KS06]). The goal of this modification of AMGe is to combine the favorable properties of classical AMG [RS87], such as an inexpensive set-up phase due to a simple coarsening procedure based on strong connections, with the superior convergence properties of AMGe, which are achieved by local harmonic interpolation.

A basic step in the construction of this method, is the computation of so-called "edge matrices", which represent the nodal dependence, and, when assembled globally, define a spectrally equivalent auxiliary problem. This auxiliary problem determines the coarsening process. Our approach originates in so-called element preconditioning techniques first introduced in [HLRS01, LRS00]. The computation of edge matrices has initially been considered in [KS06] for scalar problems, and has then been generalized to problems in linear elasticity in Reference [Kra08].

In this chapter we offer an alternative way to compute edge matrices in linear elasticity that improves the approximation properties of the auxiliary problem. Moreover, we propose a natural measure for the strength of nodal dependence defined via the constant in the strengthened Cauchy-Bunyakowski-Schwarz inequality associated with local (vertex) subspaces, and provide a two-level convergence analysis of the obtained AMGm method. Furthermore, we comment on parallelization aspects of the presented method. Towards the end of the chapter, in Section 4.8 we present several numerical experiments summarizing convergence results for reference configurations of three-dimensional bodies, including the cases of orthotropic materials, e.g., cancellous bone, hard wood, or soft wood, and problems with jumps in the Young's modulus of elasticity. Additionally, we present an example in which we compare the modified AMGm method, as described in this chapter, with the original method in [Kra08] and also with BoomerAMG ([HMY00]). Finally, in Section 4.9 we address the application of AMGm to the stable DG discretization being presented in Subsection 2.4.3.

4.1 Simple reformulation of the linear elasticity system

The focus of this work is on problems arising in linear elasticity. Therefore, let the reference configuration of an elastic body Ω be a bounded, connected and open subset of $\mathbb{R}^d$, $d = 2, 3$. From Subsection 2.3 we know that the governing equations of our problem are given by

$$-\operatorname{div} \boldsymbol{\sigma} = \boldsymbol{f} \qquad \text{in } \Omega, \tag{4.2a}$$

$$\boldsymbol{u} = \mathbf{0} \qquad \text{on } \Gamma_D, \tag{4.2b}$$

$$\boldsymbol{\sigma} \cdot \boldsymbol{n} = \boldsymbol{t}_N \qquad \text{on } \Gamma_N. \tag{4.2c}$$

These equations describe the deformation of the body under the influence of body and surface forces $\boldsymbol{f}$ and $\boldsymbol{t}_N$, respectively. The Dirichlet boundary conditions $\boldsymbol{u} = \boldsymbol{u}_D = \mathbf{0}$ on Γ_D guarantee the uniqueness of the solution if $\operatorname{meas}(\Gamma_D) > 0$.

We focus on linear elastic materials and hence the stress-strain relation by Hooke's law (2.43). First of all, we consider St. Vernant-Kirchhoff-materials (homogeneous and isotropic), whereas the stress-strain matrix C_{iso} is defined by (2.47). In this case (4.2a) yields the classical Lamé differential equation

$$-2\mu \operatorname{div} \boldsymbol{\varepsilon}(\boldsymbol{u}) - \lambda \operatorname{grad} \operatorname{div} \boldsymbol{u} = \boldsymbol{f}, \tag{4.3}$$

based on which we derive the following weak formulation of the boundary-value problem (4.2): Find $\boldsymbol{u} \in \boldsymbol{V} := [H^1_{0,\Gamma_D}(\Omega)]^d$ such that

$$a(\boldsymbol{u}, \boldsymbol{v}) = L(\boldsymbol{v}) \qquad \forall \boldsymbol{v} \in \boldsymbol{V} \tag{4.4}$$

with $a(.,.)$ defined by (2.49) where $C = C_{\text{iso}}$. Alternatively, this can be written as

$$a_{\text{iso}}(\boldsymbol{u}, \boldsymbol{v}) := a(\boldsymbol{u}, \boldsymbol{v}) \quad = \quad 2\mu(\boldsymbol{\varepsilon}(\boldsymbol{u}), \boldsymbol{\varepsilon}(\boldsymbol{v})) + \lambda (\operatorname{div} \boldsymbol{u}, \operatorname{div} \boldsymbol{v}) . \tag{4.5}$$

The right-hand side $L(\boldsymbol{v})$ is given by (2.50), where $\boldsymbol{f} \in [L_2(\Omega)]^d$ and $\boldsymbol{t}_N \in [L_2(\Gamma_N)]^d$.

To investigate the robustness of the proposed method we also consider orthotropic materials (cf. [Kik86, YKvR+99]). In this case Hooke's law is given by $\boldsymbol{\sigma} = C_{\text{ortho}} \cdot \boldsymbol{\varepsilon}$ where C_{ortho} is defined by (2.48). Hence, using (2.48) in (4.2a) yields the bilinear form (2.49) with $C = C_{\text{ortho}}$, i.e.,

$$a_{\text{ortho}}(\boldsymbol{u}, \boldsymbol{v}) := \int_\Omega C_{\text{ortho}} \cdot \boldsymbol{\varepsilon}(\boldsymbol{u}) : \boldsymbol{\varepsilon}(\boldsymbol{v}) \, d\boldsymbol{x} . \tag{4.6}$$

The existence and uniqueness of the solution of the variational problem (2.24) for both bilinear forms, (4.5) and (4.6), is discussed in Section 2.4.1

In order to solve this system numerically we use finite elements. Therefore, we consider a shape-regular triangulation $\mathcal{T}_h = \{T\}$ of the d-dimensional (here $d = 3$) domain Ω. We use tetrahedral meshes. Furthermore, we restrict all our considerations to first-order schemes in this chapter. That is, we use the space $\boldsymbol{V}_h := [V_h^1]^d$.

We focus on linear systems (4.1) that stem from the following finite element problem.

Problem 4.1.1. *Let us consider problem* (2.33), *where the bilinear form* $a(\cdot,\cdot)$ *is either given by* (4.5) *or by* (4.6) *and the linear form* $L(\cdot)$ *is defined by* (2.50).

When employing the bilinearform $a_{\text{iso}}(\cdot,\cdot)$ and $a_{\text{ortho}}(\cdot,\cdot)$ we refer to the problem as isotropic or orthotropic (linear elasticity) problem, respectively.

4.2 Approximation via edge matrices

The AMGm method, introduced in [KS06] for scalar problems and extended to problems in linear elasticity in [Kra08], is based on the construction of edge matrices, which can be used for measuring the nodal dependence. In the following we briefly describe this approach thereby taking it to a more abstract level.

We therefore introduce the terms of *algebraic vertices* and *algebraic edges*. Let $\mathcal{D} := \{d_i : i \in \{1, \ldots, n\}\}$ be the set of n degrees of freedom (DOF).

Definition 4.2.1 (algebraic vertex, algebraic edge and edge matrix). *An* algebraic vertex $\boldsymbol{v}_i$ *is an accumulation of n_{vd} vertex degrees of freedom (VDOF), i.e.,* $\boldsymbol{v}_i = \{d_{i_j} : j \in \{1, \ldots, n_{vd}\}\}$. *Furthermore for any* $\boldsymbol{v}_i$ *and* $\boldsymbol{v}_j$ *with* $i \neq j$ *it holds that* $\boldsymbol{v}_i \cap \boldsymbol{v}_j = \emptyset$. *Let* $\mathcal{V}$ *denote the set of all vertices and* n_v *its cardinality, that is* $n_v = |\mathcal{V}|$.
An algebraic edge e_{ij} *is the union of VDOF of a pair of vertices* $\boldsymbol{v}_i$ *and* $\boldsymbol{v}_j$. *Moreover,* $\mathcal{E} := \{e_{ij} : 1 \leq i < j < n_v\}$ *denotes the set of all* n_E *edges*[1].
An edge matrix E_{ij} *is an* $2n_{vd} \times 2n_{vd}$*-matrix associated with the edge* e_{ij} *which represents the relation between the algebraic vertices* $\boldsymbol{v}_i$ *and* $\boldsymbol{v}_j$[2].

Definition 4.2.1 is rather general and allows us to extend the following concept to different discretizations or finite elements, respectively. In our particular situation, we identify the algebraic vertices with the nodes of the mesh and assume that $n_{vd} = d$. Since we restrict ourselves to continuous, piecewise linear shape functions this choice is quite natural.

Aligning with Definition 4.2.1 we define the set of neighbors $\mathcal{N}_i$ of a vertex $\boldsymbol{v}_i$, that is $\mathcal{N}_i := \{\boldsymbol{v}_j : e_{ij} \in \mathcal{E}\}$.

Our aim is to determine the nodal dependence by means of edge matrices. For this sake we first have to investigate their desired properties. Following [Kra08], let $\mathcal{A} = \{A_T : T \in \mathcal{T}_h\}$ bet the set of element matrices A_T corresponding to our FE discretization of our problem based on the triangulation $\mathcal{T}_h$ of the d-dimensional domain Ω. From the coercivity of the bilinearform $a(\cdot, \cdot)$ it follows that $A_T \geq 0$, that is, A_T is symmetric positive semidefinite (SPSD).

We target in approximating each single element matrix A_T by a sum of corresponding edge matrices E_{ij}. That is, we want a set of edge matrices $\mathcal{B}_T$ associated with the set of algebraic edges $\mathcal{E}_T$ of T, such that $A_T \approx B_T := \sum_{E_{ij} \in \mathcal{B}_T} E_{ij}$. Here summation has to be understood

[1] Note that we do not distinguish between e_{ij} and e_{ji} in general since we are dealing with symmetric problems.
[2] Aligning with the convention of the edges we have $E_{ji} := E_{ij}$ for $i < j$.

in the sense of assembling. Note that $\mathcal{B}_T$ contains only one matrix E_{ij} corresponding to each edge $e_{ij} \in \mathcal{E}_T$. This setting aligns with our general adjustment to set up one edge matrix for each algebraic edge. We want to compute an approximation B_T of A_T via a set of SPSD matrices $\mathcal{B}_T$ such that B_T is spectrally as close as possible to A_T. This idea has already been considered in [HLRS01]. For a more detailed discussion for systems of equations as in our case see [Kra08].

In the following part of this section, we characterize such approximate splittings. First, we note that an approximate splitting yields a finite (effective) condition number, i.e.

$$\kappa(A_T,\, B_T) := \frac{\inf\left\{\lambda\,:\, \boldsymbol{x}^T A_T \boldsymbol{x} \le \lambda \boldsymbol{x}^T B_T \boldsymbol{x} \;\; \forall \boldsymbol{x} \in \mathbb{R}^d\right\}}{\sup\left\{\lambda\,:\, \boldsymbol{x}^T A_T \boldsymbol{x} \ge \lambda \boldsymbol{x}^T B_T \boldsymbol{x} \;\; \forall \boldsymbol{x} \in \mathbb{R}^d\right\}} < \infty\,, \tag{4.7}$$

if the splitting $\mathcal{B}_T$ is *kernel preserving*. In other words, for every splitting $\mathcal{B}_T$ with SPSD edge matrices $E_{ij} \in \mathcal{B}_T$, we have

$$\kappa(A_T,\, B_T) < \infty \;\Leftrightarrow\; \ker(B_T) = \ker(A_T) \;\Rightarrow\; \ker(E_{ij}) = \ker(A_T)|_{(ij)} \quad \forall E_{ij} \in \mathcal{B}_T \tag{4.8}$$

For a proof see [Kra08]. Being aware of the necessity of kernel preservation, one can investigate the required properties of the edge matrices. In [Kra08] the following theorem has been shown.

Theorem 4.2.2 (characterization of semipositive splittings). *Let A_T be an SPSD element matrix arising form the (first-order) FE discretization of the 3D elasticity problem, Problem 4.1.1, using a (shape-regular) tetrahedral triangulation $\mathcal{T}_h$. Further, let $\mathcal{B}_T = \{E_{ij} \,:\, E_{ij} \ge 0 \,\wedge\, 1 \le i < j \le 4\}$ be a set of SPSD matrices providing the splitting $B_T = \sum_{E_{ij} \in \mathcal{B}_T} E_{ij}$.*
Then $\kappa(A_T,\, B_T) < \infty$ if and only if the $2n_{vd} \times 2n_{vd}$ (nonzero) edge matrices E_{ij} have the form

$$E_{ij} = c_{ij} \begin{pmatrix} \boldsymbol{v}_{ij}^T \boldsymbol{v}_{ij} & -\boldsymbol{v}_{ij}^T \boldsymbol{v}_{ij} \\ -\boldsymbol{v}_{ij}^T \boldsymbol{v}_{ij} & \boldsymbol{v}_{ij}^T \boldsymbol{v}_{ij} \end{pmatrix} \quad \text{with} \quad \boldsymbol{v}_{ij} = \begin{pmatrix} x_j - x_i \\ y_j - y_i \\ z_j - z_i \end{pmatrix} \qquad 1 \le i < j \le 4, \tag{4.9}$$

and $c_{ij} > 0$, where $(x_k,\, y_k,\, z_k)$, $1 \le k \le 4$, denote the vertices of the tetrahedron T.

It is obvious that the edge matrices in Theorem 4.2.2 are of rank one. Additionally the exact representation of each E_{ij} is given by (4.9), which is determined up to the scalar constant c_{ij} since the vector $\boldsymbol{v}_{ij}$ is the directional vector of the edge e_{ij}.

As compared to the method presented in [Kra08] we modify the construction of the edge matrices E_{ij}. This modification has several advantages, e.g., the computation of the constants c_{ij} is simple and computationally inexpensive. Let us consider the matrix G_h, which is assembled from all element stiffness matrices A_T of our problem. That is, we neglect all boundary terms. Hence, we have

$$A = A_h = \sum_{T \in \mathcal{T}_h} A_T + F_h =: G_h + F_h \,, \tag{4.10}$$

where F_h contains the boundary terms. The matrix G_h can be represented in the form

$$G_h = \begin{bmatrix} \ddots & \vdots & \vdots & \vdots & \vdots & \vdots & \vdots \\ \dots & G_{ij} + G_{ik} + \dots & \dots & -G_{ij} & \dots & -G_{ik} & \dots \\ \vdots & \vdots & \ddots & \vdots & \vdots & \vdots & \vdots \\ \dots & -G_{ij}^T & \dots & G_{ij}^T + \dots & \dots & \dots & \dots \\ \vdots & \vdots & \vdots & \vdots & \ddots & \vdots & \vdots \\ \dots & -G_{ik}^T & \dots & \dots & \dots & G_{ik}^T + \dots & \dots \\ \vdots & \vdots & \vdots & \vdots & \vdots & \vdots & \ddots \end{bmatrix} \tag{4.11}$$

with $n_v \times n_v$ blocks of size $n_{vd} \times n_{vd}$. Based on the off-diagonal blocks of G_h we want to determine the contribution of the edge connecting the vertices $\boldsymbol{v}_i$ and $\boldsymbol{v}_j$ (algebraic edge e_{ij}). One possible choice is to use

$$\tilde{E}_{ij}^{(1)} = \begin{pmatrix} G_{ij} & -G_{ij} \\ -G_{ij}^T & G_{ij}^T \end{pmatrix}, \tag{4.12}$$

which is a unsymmetric matrix in general because G_{ij} is not necessarily symmetric. On the other hand, since the diagonal blocks of G_h are symmetric $\tilde{E}_{ij}$ can also be chosen according to

$$\tilde{E}_{ij}^{(2)} = \begin{pmatrix} \frac{1}{2}(G_{ij} + G_{ij}^T) & -G_{ij} \\ -G_{ij}^T & \frac{1}{2}(G_{ij} + G_{ij}^T) \end{pmatrix}. \tag{4.13}$$

Note that $\tilde{E}_{ij}^{(2)}$ is symmetric. Now, we aim at approximating $\tilde{E}_{ij}^{(p)}$, $p = 1, 2$, by

$$E_{ij} := \arg\min_E \left\| E - \tilde{E}_{ij}^{(p)} \right\|^2 . \tag{4.14}$$

Since we know the form of E_{ij} (cf. Theorem 4.2.2) we only have to determine c_{ij}. The solution of (4.14) is much easier when using the Frobenius norm instead of the l_2-norm. With this choice we arrive at

$$c_{ij} = \arg\min_c \left\| E_{ij}(c) - \tilde{E}_{ij}^{(p)} \right\|_F^2 \tag{4.15}$$

with

$$E_{ij}(c) = c \begin{pmatrix} \boldsymbol{v}_{ij}^T \boldsymbol{v}_{ij} & -\boldsymbol{v}_{ij}^T \boldsymbol{v}_{ij} \\ -\boldsymbol{v}_{ij}^T \boldsymbol{v}_{ij} & \boldsymbol{v}_{ij}^T \boldsymbol{v}_{ij} \end{pmatrix},$$

where $\boldsymbol{v}_{ij}$ is the directional vector of the edge e_{ij}.

Theorem 4.2.3. *Let $\tilde{E}_{ij}^{(p)}$, $p = 1, 2$, be defined by (4.12) or (4.13). Then the minimization problem (4.15) has the solution*

$$c_{ij} = \frac{\boldsymbol{v}_{ij}^T G_{ij} \boldsymbol{v}_{ij}}{\|\boldsymbol{v}_{ij}\|_2^4} . \tag{4.16}$$

Proof. For simplicity we neglect the subscripts ij of the occurring vectors and matrices. From the representation of the Frobenius norm via the trace-operator we obtain

$$\|E(c) - \tilde{E}^{(p)}\|_F^2 = \operatorname{tr}(E(c)^2) + \operatorname{tr}((\tilde{E}^{(p)})^2) - 2\operatorname{tr}(E(c)\,\tilde{E}^{(p)}) . \tag{4.17}$$

Thus we have

$$\operatorname{tr}(E(c)^2) = \operatorname{tr}\left(c^2 \begin{bmatrix} 2\boldsymbol{v}\boldsymbol{v}^T\boldsymbol{v}\boldsymbol{v}^T & -2\boldsymbol{v}\boldsymbol{v}^T\boldsymbol{v}\boldsymbol{v}^T \\ -2\boldsymbol{v}\boldsymbol{v}^T\boldsymbol{v}\boldsymbol{v}^T & 2\boldsymbol{v}\boldsymbol{v}^T\boldsymbol{v}\boldsymbol{v}^T \end{bmatrix}\right) = 4c^2\|\boldsymbol{v}\|_2^4 , \tag{4.18}$$

and (for $p = 1$)

$$\begin{aligned} \operatorname{tr}(E(c)\,\tilde{E}^{(1)}) &= \operatorname{tr}\left(c \begin{bmatrix} \boldsymbol{v}\boldsymbol{v}^T(G+G^T) & -\boldsymbol{v}\boldsymbol{v}^T(G+G^T) \\ -\boldsymbol{v}\boldsymbol{v}^T(G+G^T) & \boldsymbol{v}\boldsymbol{v}^T(G+G^T) \end{bmatrix}\right) \\ &= 2c\operatorname{tr}(\boldsymbol{v}\boldsymbol{v}^T(G+G^T)) = 2c(\operatorname{tr}(\boldsymbol{v}\boldsymbol{v}^T G) + \operatorname{tr}(\boldsymbol{v}\boldsymbol{v}^T G^T)) \\ &= 4c\boldsymbol{v}^T G\boldsymbol{v} . \end{aligned} \tag{4.19}$$

For $p = 2$ the matrix product looks different, but its trace is identical to (4.19). In summary, by plugging (4.18) and (4.19) into (4.17) we obtain

$$\zeta(c) := \|E(c) - \tilde{E}^{(p)}\|_F^2 = 4c^2\|\boldsymbol{v}\|_2^4 + \operatorname{tr}((\tilde{E}^{(p)})^2) - 8c\boldsymbol{v}^T G\boldsymbol{v} .$$

Then, solving $\zeta'(c) = 0$ for c results in (4.16). Since $\zeta''(c)$ is positive, the solution is a unique minimum. □

Considering Problem 4.1.1 it is possible that the off-diagonal submatrices G_{ij} of the matrix G_h are indefinite and hence it might happen that (4.16) yields $c_{ij} < 0$. In this case we set $c_{ij} = -c_{ij}$. Finally, note that we have $c_{ij} \geq 0$ if G_{ij} is positive semidefinite which is the usual case.

4.3 Detection of strong couplings (nodal dependence)

In any multigrid method the error is reduced by relaxation on the one hand and by coarse-grid correction on the other hand. These two components should complement each other which means that error modes not effected by one of these two components should be treated efficiently by the other component. In AMG one first chooses the smoother M, e.g., Gauss-Seidel relaxation, and then constructs a coarse-grid correction that is capable of reducing algebraically smooth error modes, namely errors $\boldsymbol{e}$ for which

$$||S\boldsymbol{e}||_A \approx ||\boldsymbol{e}||_A \tag{4.20}$$

holds (cf. [RS87]), where $S = I - M^{-1}A$ is the error propagation matrix (iteration matrix) of the relaxation process. This leads, for Gauss-Seidel-smoothers among others, to the condition that the residual $\boldsymbol{r} = A\boldsymbol{e}$ has to be small as compared to the error itself, which means that

$$(r_i =) a_{ii}e_i + \sum a_{ij}e_j \approx 0\,. \tag{4.21}$$

In other words, the i-th error-component e_i is mainly determined by those e_j for which the corresponding $|a_{ij}|$ are large. In standard AMG methods (cf. [RS87, CFH+98, TOS01]), the coarse grid is selected based on a measure for the strength of nodal dependence. According to Reference [RS87] point i *strongly depends* on j if

$$-a_{ij} \geq \theta \max_{l \neq i}\{-a_{il}\}\,,$$

with some threshold $\theta \in (0, 1]$, e.g., $\theta = 0.25$. This concept works very well for M-matrices (and for small perturbations of M-matrices), but for non-M-matrices it is not clear how to take into account positive and negative off-diagonal entries. As a remedy, proposed in [CV00], for finite element problems, strong connections (i, j) can be selected based on the criterion

$$\frac{|a_{ij}|}{\sqrt{a_{ii}\, a_{jj}}} \geq \theta\,. \tag{4.22}$$

Thereby, the entries of the local stiffness matrices A_T are used. Note that (4.22) measures the energy-cosine of the abstract angle between the i-th and the j-th nodal basis function. While this approach has no difficulties with non-M-matrices, it is not directly applicable in the multilevel setting because the element stiffness matrices are usually not available on all levels. Another disadvantage of measure (4.22) is that it is not obvious what its generalization is for vector-field problems. Considering each DOF for its own typically yields more connections than desired and does not result in a reliable measure in general (cf. [TOS01, Cho01]).

Therefore, we propose the following approach to measure the strength of an algebraic edge e_{ij} connecting the algebraic vertices $\boldsymbol{v}_i$ and $\boldsymbol{v}_j$. We collect all edges e_{kl}, i.e., edge matrices E_{kl}, building a loop of length 3 which contains the edge e_{ij}. By assembling all these matrices E_{kl} we get an SPSD so-called *computational molecule* $M(i, j)$.

$$M(i,\, j) = E_{ij} + \sum_{k \in \mathcal{N}_i \cap \mathcal{N}_j} (E_{ik} + E_{jk}) = \begin{bmatrix} M_{ii} & M_{ij} & M_{ik} \\ M_{ji} & M_{jj} & M_{jk} \\ M_{ki} & M_{kj} & M_{kk} \end{bmatrix} \tag{4.23}$$

with $M_{ii},\, M_{jj} \in \mathbb{R}^{n_{vd} \times n_{vd}}$ and $M_{kk} \in \mathbb{R}^{|\mathcal{N}_i \cap \mathcal{N}_j| n_{vd} \times |\mathcal{N}_i \cap \mathcal{N}_j| n_{vd}}$. Now, we consider the submatrix of (4.23) corresponding to the unknowns of $\boldsymbol{v}_i$ and $\boldsymbol{v}_j$ based on which we shall measure the cosine of the angle between the subspaces related to $\boldsymbol{v}_i$ and $\boldsymbol{v}_j$, which is given by the smallest possible constant in the strengthened *Cauchy-Bunyakowski-Schwarz inequality* (CBS constant).

Definition 4.3.1 (strong connection via CBS constant). *Let the computational molecule be defined as in* (4.23). *Then the strength* s_{ij} *of the algebraic edge* e_{ij} *is defined by*

$$s_{ij} := \sqrt{1 - \inf_{\boldsymbol{x}} \frac{\boldsymbol{x}^t S_{ii} \boldsymbol{x}}{\boldsymbol{x}^t M_{ii} \boldsymbol{x}}}\,, \tag{4.24}$$

where $S_{ii} := M_{ii} - M_{ij} M_{jj}^{-1} M_{ji}$.

Note that in Definition 4.3.1 we implicitly assume that M_{jj} is regular. If this is not the case, the Schur complement S_{ii} is replaced by the *generalized Schur complement* introduced in [Kra08] (Algorithm 3.9). Nevertheless, if M_{ii} and M_{jj} both are regular, the strength s_{ij} of the edge e_{ij} is independent of the order of i and j and thus s_{ij} can equivalently be computed according to

$$s_{ij} = \sqrt{1 - \inf_{\boldsymbol{x}} \frac{\boldsymbol{x}^t S_{jj} \boldsymbol{x}}{\boldsymbol{x}^t M_{jj} \boldsymbol{x}}}\,,$$

with the Schur complement $S_{jj} := M_{jj} - M_{ji} M_{ii}^{-1} M_{ij}$. This follows directly from the definition of the CBS constant (cf. [Axe94]).

Now, according to (4.22), we say a connection between the vertices $\boldsymbol{v}_i$ and $\boldsymbol{v}_j$ is strong, $\boldsymbol{v}_i$ strongly depends on $\boldsymbol{v}_j$, or the edge e_{ij} is strong, iff

$$s_{ij} \geq \theta\,, \tag{4.25}$$

with $\theta \in (0,\, 1]$. In other words, a connection between two algebraic vertices is strong if the angle between their respective subspaces, generated by $M(i, j)$, is less than a certain threshold $\arccos(\theta)$. In the following, let $\mathcal{S}_i$ denote the set of strongly connected neighbors of a vertex $\boldsymbol{v}_i$, i.e., $\mathcal{S}_i := \{\boldsymbol{v}_j \; : \; \boldsymbol{v}_j \in \mathcal{N}_i \wedge s_{ij} \geq \theta\}$.

It is reasonable to choose $\theta \in (0, 1]$ in order to control the fraction $\Theta \cdot n_E$ of weak edges. Later, in the numerical examples (see Section 4.8) we choose $\Theta = 0.08$ (or $\Theta = 0.06$ in Subsection 4.8.2). These settings have shown to be appropriate in case of the considered examples. Note that increasing the number of strong edges, i.e., decreasing Θ, typically results in a faster coarsening.

4.4 Coarse-grid selection

Based on the *strength of connectivity* s_{ij}, associated with the algebraic edges e_{ij}, we select our coarse grid. Following the recent articles [Kra08, KS06] the same coarsening strategy as explained in [KS06] is used, which is a slight modification of the coarse-grid selection procedure proposed by Ruge and Stüben in [RS87].

For the sake of completeness, we depict the procedure in this work. The selection process is guided by two criterions that can be found for instance in [KS06], or originally in [RS87]. They are

C1: $\mathcal{V}_c$ should be a maximum independent set, which means that no strong connections are allowed within $\mathcal{V}_c$.

C2: Each vertex $\boldsymbol{v}_j$ being strongly connected to a fine-vertex $\boldsymbol{v}_i$ is either contained in $\mathcal{V}_c$ or it strongly depends on at least one coarse node $\boldsymbol{v}_k$ that itself is strongly connected to vertex $\boldsymbol{v}_i$.

We partition the set of vertices $\mathcal{V}$ into a set of fine vertices $\mathcal{V}_f$, $n_{vf} := |\mathcal{V}_f|$, and a set of coarse vertices $\mathcal{V}_c$, $n_{vc} := |\mathcal{V}_c|$. Corresponding to this partitioning we define the set of fine neighbors $\mathcal{N}_i^f := \{\boldsymbol{v}_j \; : \; \boldsymbol{v}_j \in \mathcal{N}_i \cap \mathcal{V}_f\}$ and the set of coarse neighbors $\mathcal{N}_i^c := \{\boldsymbol{v}_j \; : \; \boldsymbol{v}_j \in \mathcal{N}_i \cap \mathcal{V}_c\}$ of a vertex $\boldsymbol{v}_i$. Moreover, let the sets of strongly connected fine and coarse neighbors be denoted by $\mathcal{S}_i^f := \{\boldsymbol{v}_j \; : \; \boldsymbol{v}_j \in \mathcal{S}_i \cap \mathcal{V}_f\}$ and $\mathcal{S}_i^c := \{\boldsymbol{v}_j \; : \; \boldsymbol{v}_j \in \mathcal{S}_i \cap \mathcal{V}_c\}$, respectively.

The selection process is executed in a fast two-stage process. First, by a simple Greedy-algorithm, Algorithm 4.4.1, criterion (C1) is fulfilled. This is because if a vertex is selected to be a coarse node, all its strongly connected neighbors are added to the set of fine vertices $\mathcal{V}_f$.

Second, Algorithm 4.4.2 enforces (C2). It successively checks for each fine vertex $\boldsymbol{v}_i$ if all of their strongly connected fine neighbors $\boldsymbol{v}_j$ are strongly connected to a strong coarse neighbor of $\boldsymbol{v}_i$. If this is not the case, either $\boldsymbol{v}_i$ is added to $\mathcal{V}_c$ if $|\mathcal{V}_c \cap \mathcal{S}_i| < |\mathcal{V}_c \cap \mathcal{S}_j|$ or if $n_1 \leq n_2$, $\boldsymbol{v}_j$ is switched to be coarse.

Algorithm 4.4.1: Coarse-grid selection

$\mathcal{V}_f = \mathcal{V}_c = \emptyset$ and $\tilde{\mathcal{V}} = \mathcal{V}$
for $i = 1, \ldots, n_v$ **do**
  $\lambda_i = |\mathcal{S}_i|$ `// initial number of strongly connected neighbors`
$m = 0$
while $m < n_v$ **do**
  find i such that $\lambda_i = \max_{m \in \tilde{\mathcal{V}}} \lambda_m$
  $\mathcal{V}_c \leftarrow \mathcal{V}_c \cup \{\boldsymbol{v}_i\}$
  $\tilde{\mathcal{V}} \leftarrow \tilde{\mathcal{V}} \backslash \{\boldsymbol{v}_i\}$
  $m \leftarrow m + 1$
  forall the $\boldsymbol{v}_j \in \mathcal{S}_i \cap \tilde{\mathcal{V}}$ **do**
    $\mathcal{V}_f \leftarrow \mathcal{V}_f \cup \{\boldsymbol{v}_j\}$
    $\tilde{\mathcal{V}} \leftarrow \tilde{\mathcal{V}} \backslash \{\boldsymbol{v}_j\}$
    $m \leftarrow m + 1$
    forall the $\boldsymbol{v}_k \in \mathcal{S}_j \cap \tilde{\mathcal{V}}$ **do**
      $\lambda_k \leftarrow \lambda_k + 1$

When we have selected the coarse grid, we can set up a prolongation operator P and compute the coarse grid matrix A_H via the Galerkin approach, i.e., $A_H := P^T A_h P$. In the case of a two-level method, the knowledge of A_h and P is sufficient to set up the whole procedure. However, in the multilevel setting we need edges e_{ij}^c and their corresponding matrices E_{ij}^c on the coarse level. Those can be constructed in a straightforward way.

We consider the matrix G_h (see (4.11)). Via the Galerkin approach we compute $G_H = P^T G_h P$. Then, if we know the coordinates of the coarse vertices we can proceed in the following way. We construct a coarse edge e_{ij}^c connecting the coarse vertices $\boldsymbol{v}_i$ and $\boldsymbol{v}_j$ if they are strongly connected on the fine grid via at most two edges. The respective edges can be determined via the adjacency matrix product $\mathcal{A}_c = ((\mathcal{A}_{fc}^s)^T \times \mathcal{A}_{fc}^s) \vee \mathcal{A}_{cc}$. Here, $\mathcal{A}_{fc}^s$ denotes the adjacency matrix of the strong connections between fine and coarse vertices, $\mathcal{A}_{cc}$ represents the coarse-to-coarse adjacency on the fine mesh, and $\mathcal{A}_c$ is the coarse-level adjacency matrix. Alternatively, one could add coarse edges between any pair of coarse vertices $\boldsymbol{v}_i$ and $\boldsymbol{v}_j$ if there exists a *strong* path of length three between them. However, if these latter mentioned coarse edges are added the coarse-grid operators of our method typically become significantly denser.

Algorithm 4.4.2: Coarse grid refinement

```
for i = 1, ..., n_v do
    λ_i = 0                                   // initialize
m = 0
while m < n_v do
    m ← m + 1
    if v_i ∈ V_f then
        n_1 = 0
        forall the v_k ∈ S_i ∩ V_c do
            n_1 ← n_1 + 1
            λ_k = 1
        forall the v_j ∈ S_i ∩ V_f do
            n_2 = n_3 = 0
            forall the v_k ∈ S_j ∩ V_c do
                n_2 ← n_2 + 1
                if λ_k = 1 then
                    n_3 ← n_3 + 1             // v_j has a strong coarse neighbor in S_i ∩ V_c
            if n_3 < 1 then
                if n_1 < n_2 then
                    V_c ← V_c ∪ {v_i}
                    V_f ← V_f \ {v_i}
                else
                    V_c ← V_c ∪ {v_j}
                    V_f ← V_f \ {v_j}
                    n_1 ← n_1 + 1
                    λ_j = 1
        forall the v_k ∈ N_i do
            λ_k = 0
```

As can be seen later (in Section 4.5), we use a locally energy-minimizing interpolation that involves the inverse of a small-sized matrix M_{ff} (cf. (4.27)–(4.29)). The regularity of M_{ff} demands a proper selection of coarse vertices locally, which in practice can be achieved by enriching $\mathcal{V}_c$.

To illustrate the effect of the coarsening process let us consider Problem 4.1.1 on the unit square Ω in 2D. We examine the effect of varying the stress-strain matrix C and also the influence of changing the mesh.

Figure 4.1(a) shows the fine grid and the first four coarse grids for an isotropic constitutive law when originating from a uniform mesh. As is reasonable the coarsening ratio is approximately the same in each direction. In Figure 4.1(b) we see the desired semi-coarsening for an orthotropic material with a different Young's modulus in x- and y-direction, namely $E_y/E_x = 30$. As Figure 4.1(c) shows, the situation is similar when using an unstructured mesh, which, however, results in a smaller grid complexity σ^Ω in this example.

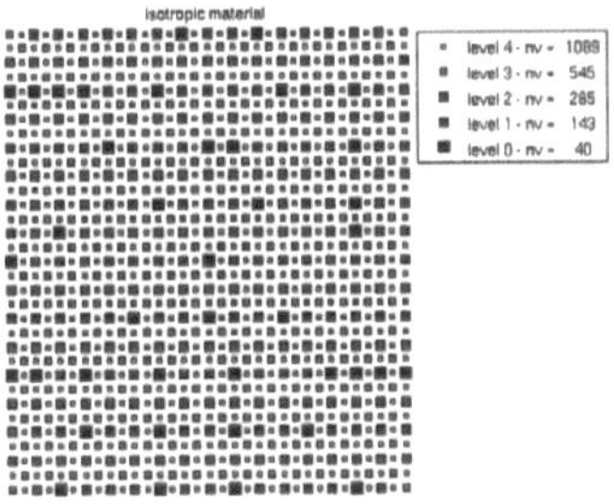

(a)

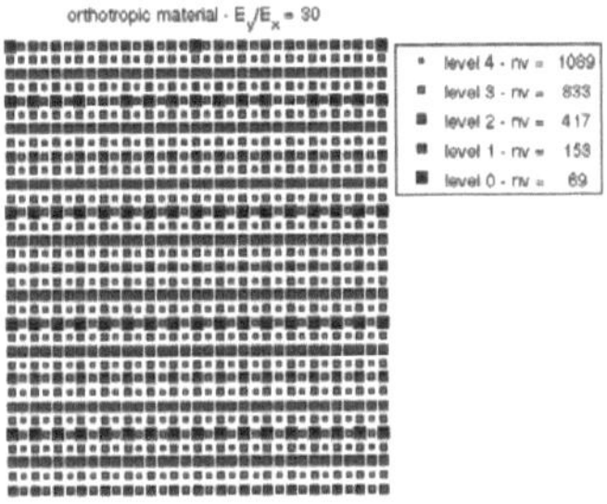

(b)

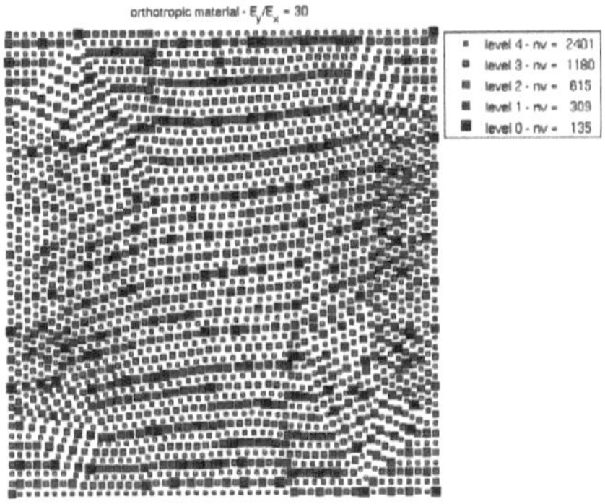

(c)

Figure 4.1: The coarse grid for different types of the stress-strain matrix on different meshes: isotropic material on a uniform gird (upper left); orthotropic material with $E_y/E_x = 30$ on a uniform mesh (upper right); unstructured mesh (lower picture)

4.5 Interpolation and smoothing

The interpolation we are using here does not differ from the one used in References [Kra08, KS06]. After reordering the vertices such that we have first the fine and then the coarse vertices, the prolongation operator $P \in \mathbb{R}^{(n_v n_{vd})\times(n_{vc} n_{vd})}$ is given by

$$P = \begin{bmatrix} P_{\mathrm{fc}} \\ I_{\mathrm{c}} \end{bmatrix}, \tag{4.26}$$

where I_{c} is the identity corresponding to the coarse vertices and P_{fc} is a mapping from the $(n_{vc} n_{vd})$-dimensional coarse space onto the $(n_{vf} n_{vd})$-dimensional fine space. In order to set up the interpolation for the DOF of a fine vertex $\boldsymbol{v}_i$, we define the so-called interpolation molecule

$$M(i) := \sum_{\boldsymbol{v}_k \in \mathcal{S}_i^c} E_{ik} + \sum_{\boldsymbol{v}_j \in \mathcal{S}_i^f : \exists \boldsymbol{v}_k \in \mathcal{S}_i^c \cap \mathcal{S}_j^c} E_{ij} + \sum_{\boldsymbol{v}_k \in \mathcal{S}_i^c, \boldsymbol{v}_j \in \mathcal{S}_i^f : \boldsymbol{v}_k \in \mathcal{S}_j^c} E_{kj} \,. \tag{4.27}$$

The molecule $M(i)$ emerges from the assembling of different edge matrices. The first sum in

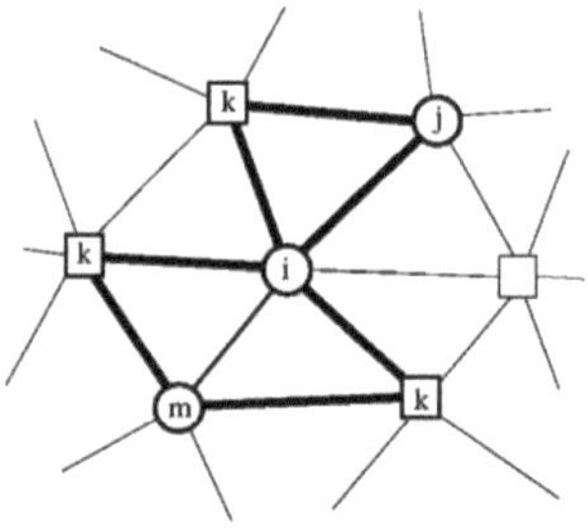

Figure 4.2: Sketch, visualizing the chosen connections for setting up $M(i)$ according to (4.26).

(4.27) comprises all edges connecting $\boldsymbol{v}_i$ strongly to a coarse vertex $\boldsymbol{v}_k$. Further, all strong fine edges e_{ij} are added if there exists a strong connection from $\boldsymbol{v}_j$ to one strong coarse neighbor $\boldsymbol{v}_k$ of $\boldsymbol{v}_i$. Additionally, all connections between the strong fine neighbors $\boldsymbol{v}_j \in \mathcal{S}_i^f$ and the strong coarse neighbors $\boldsymbol{v}_k \in \mathcal{S}_i^c$ are taken into account. In Figure 4.2, a sketch of the selection process in (4.26) is provided. Thereby, the circles denote fine vertices while squares designate coarse algebraic vertices. Thick lines represent strong connections. Therefore, the coarse vertices $\boldsymbol{v}_k$ correspond to the first part of (4.26). The connection between the fine nodes $\boldsymbol{v}_i$ and $\boldsymbol{v}_j$ is related to the second part, while the edge connecting $\boldsymbol{v}_j$ and $\boldsymbol{v}_k$ belongs to the third sum in (4.26). Contrary, the fine vertex $\boldsymbol{v}_m$ is not chosen since its connection to $\boldsymbol{v}_i$ is not strong. Hence, $M(i)$ has the following block-structure

$$M(i) = \begin{bmatrix} M_{\mathrm{ff}} & M_{\mathrm{fc}} \\ M_{\mathrm{cf}} & M_{\mathrm{cc}} \end{bmatrix}. \tag{4.28}$$

Moreover, $M(i)$ is SPSD by construction. Following the ideas of AMGe (cf. [BCF+01, FV04, KV06]) the optimal interpolation coefficients $P^*_{fc,i}$ with respect to the molecule $M(i)$ are given by

$$P^*_{\mathrm{fc,i}} := -(M_{\mathrm{ff}}^{-1} M_{\mathrm{fc}})_i \,, \tag{4.29}$$

where $P^*_{\mathrm{fc,i}}$ denotes the n_{vd} rows of the local interpolation matrix P_{fc} corresponding to the fine vertex $\boldsymbol{v}_i$. Note that this choice minimizes the following measure for the defect $\boldsymbol{d}_{\mathrm{f}} := \boldsymbol{e}_{\mathrm{f}} - P_{\mathrm{fc}}\boldsymbol{e}_{\mathrm{c}}$ of the local interpolation, i.e.,

$$P^*_{\mathrm{fc}} := \arg\min_{P} \max_{\boldsymbol{e}\perp\ker(M)} \frac{(\boldsymbol{e}_{\mathrm{f}} - P\boldsymbol{e}_{\mathrm{c}})^T(\boldsymbol{e}_{\mathrm{f}} - P\boldsymbol{e}_{\mathrm{c}})}{\boldsymbol{e}^T M \boldsymbol{e}} \,. \tag{4.30}$$

For further details we refer to [FV04, KS06]. Additionally note the similarity of the above measure to the measure μ, (3.40), which is used to quantify the condition number $\kappa(B^{-1}_{\mathrm{MG,2}}A)$.

So far, we have collected all ingredients that are required to set up an AMG algorithm. As it is quite common, we use AMG to precondition the conjugate gradient (CG) method. Our choice of the smoothing operator S is Gauss-Seidel iteration with a certain number of pre- and post-smoothing steps ν_1 and ν_2. The 2-grid algorithm is depicted in Algorithm 3.2.1. with the error propagation $\boldsymbol{e}^{(k+1)} = (I - B^{-1}_{\mathrm{AMG,2}}A)\boldsymbol{e}^{(k)}$. On the other hand, for higher n_L, we arrive at he multigrid algorithm 4, which defines the preconditioner B^{-1}_{AMG}.

4.6 Two-level convergence

In this section we investigate the convergence properties of the two-level AMGm method, see Algorithm 3.2.1. As mentioned in the previous section we use AMG as a preconditioner for CG and hence $\kappa(B^{-1}_{\mathrm{AMG,2}}A)$, is the decisive measure for the convergence rate. The derivations in this section are based on Section 3.3 and on the results in References [Not05] and [FV04].

Let us reconsider the setting of Section 3.3. Especially, the splitting (3.35) is used and we exploit the results of Theorem 3.3.1. In the following, we examine the CBS constant $\hat{\gamma}$ for our method AMGm. Therefore, we use the spectral equivalence of the element matrices A_T and the molecule matrices B_T assembled from the set $\mathcal{B}_T$ (see Section 4.2). That is, for each element $T \in \mathcal{T}_h$, there exist constants $\underline{c}_T$, $\overline{c}_T > 0$, such that

$$\underline{c}_T(B_T\boldsymbol{v}\,,\,\boldsymbol{v}) \le (A_T\boldsymbol{v}\,,\,\boldsymbol{v}) \le \overline{c}_T(B_T\boldsymbol{v}\,,\,\boldsymbol{v}) \qquad \forall \boldsymbol{v}\,. \tag{4.31}$$

Summing up the contributions $E_{ij,T}$ of all elements T that share a given edge e_{ij}, finally yields the edge matrix E_{ij}, i.e., $E_{ij} = \sum_{T\,:\,e_{ij}\in\mathcal{E}_T} E_{ij,T}$. Using $G = \sum_{T\in\mathcal{T}_h} A_T$ (see (4.10)) and

$B := \sum_{e_{ij}\in\mathcal{E}} E_{ij} = \sum_{T\in\mathcal{T}_h} B_T$ gives

$$\underline{c}(B\boldsymbol{v}\,,\,\boldsymbol{v}) \le (G\boldsymbol{v}\,,\,\boldsymbol{v}) \le \overline{c}(B\boldsymbol{v}\,,\,\boldsymbol{v}) \qquad \forall \boldsymbol{v} \tag{4.32}$$

with $\underline{c} := \min_{T\in\mathcal{T}_h} \underline{c}_T$ and $\overline{c} := \max_{T\in\mathcal{T}_h} \overline{c}_T$.

Now, let us examine the prolongation we are using according to Section 4.5. For simplicity in the following we assume that for interpolation of a fine vertex $\boldsymbol{v}_i \in \mathcal{V}_f$ the computational molecule consists of all coarse-neighbor connections of vertex $\boldsymbol{v}_i$, i.e., $M(i) := \sum_{\boldsymbol{v}_j\in\mathcal{N}_i^c} E_{ij}$. Then the matrix B can be rewritten in the form

$$B = \sum_{\boldsymbol{v}_i\in\mathcal{V}_f} M(i) + \sum_{\boldsymbol{v}_i,\,\boldsymbol{v}_j\in\mathcal{V}_f\wedge e_{ij}\in\mathcal{E}} E_{ij} + \sum_{\boldsymbol{v}_i,\,\boldsymbol{v}_j\in\mathcal{V}_c\wedge e_{ij}\in\mathcal{E}} E_{ij} =: M + H\,, \tag{4.33}$$

where

$$M := \sum_{\boldsymbol{v}_i\in\mathcal{V}_f} M(i) \quad \text{and} \quad H := B - M = \begin{bmatrix} H_{\mathrm{ff}} & 0 \\ 0 & H_{\mathrm{cc}} \end{bmatrix}.$$

Hence, B is the sum of all interpolation molecules M plus a block-diagonal remainder H. In the following we estimate the CBS constant $\gamma(\hat{B})$ of the matrix $\hat{B} := J^T B J$. Since $J^T M(i) J$ is a block diagonal matrix[3] for all $\boldsymbol{v}_i \in \mathcal{V}_f$ it follows that $\gamma(J^T M(i) J) = 0$. Thus $\gamma(J^T M J) = 0$, cf. Lemma 5.1 in [LM07]. For the examination of $\gamma(\hat{H}) = \gamma(J^T H J)$ we write $\hat{B} = J^T B J$ in the form

$$\hat{B} = \hat{M} - c_0 T + \hat{H} + c_0 T$$

with $c_0 > 0$ and the block-diagonal matrix $T := \operatorname{diag}(0, P_{\mathrm{fc}}^T H_{\mathrm{ff}} P_{\mathrm{fc}})$. If $M_{\mathrm{cc}} - c_0 P_{\mathrm{fc}}^T H_{\mathrm{ff}} P_{\mathrm{fc}}$ is SPSD and does not vanish, we have $\gamma(\hat{M} - c_0 T) = 0$. These requirements are fulfilled for sufficiently small $c_0 > 0$ due to the construction of P_{fc}.

On the other hand, with J given by (3.38), we get

$$\bar{H} := \hat{H} + c_0 T = J^T H J + c_0 T = \begin{bmatrix} H_{\mathrm{ff}} & H_{\mathrm{ff}} P_{\mathrm{fc}} \\ P_{\mathrm{fc}}^T H_{\mathrm{ff}} & H_{\mathrm{cc}} + (1 + c_0) P_{\mathrm{fc}}^T H_{\mathrm{ff}} P_{\mathrm{fc}} \end{bmatrix}.$$

Since H_{ff} and H_{cc} are SPSD we have $(H_{\mathrm{cc}} + (1+c_0) P_{\mathrm{fc}}^T H_{\mathrm{ff}} P_{\mathrm{fc}}) \boldsymbol{v}_c = 0$ for all $(\boldsymbol{v}_f^T, \boldsymbol{v}_c^T)^T \in \ker(\bar{H})$, which implies that $\gamma(\bar{H}) < 1$, cf. Lemma 9.1 in [Axe94]. Now with the basis matrices B_{f} and B_{c} of $\ker(H_{\mathrm{ff}})^{\perp}$ and $\ker(H_{\mathrm{cc}} + (1+c_0) P_{\mathrm{fc}}^T H_{\mathrm{ff}} P_{\mathrm{fc}})^{\perp} = \ker(P_{\mathrm{fc}}^T H_{\mathrm{ff}} P_{\mathrm{fc}})^{\perp}$ we set up the l_2-orthogonal projections $P_{\mathrm{f}} = B_{\mathrm{f}} (B_{\mathrm{f}}^T B_{\mathrm{f}})^{-1} B_{\mathrm{f}}^T$ and $P_{\mathrm{c}} = B_{\mathrm{c}} (B_{\mathrm{c}}^T B_{\mathrm{c}})^{-1} B_{\mathrm{c}}^T$ (compare Subsection 3.6.1). Since

[3] This is due to the special form of P_{fc}, see (4.29).

$\gamma(\bar{H}) < 1$ we can view the matrices in terms of the bases and we arrive at

$$\gamma(P_{\bar{H}}^T \bar{H} P_{\bar{H}}) = \gamma(\bar{H}),$$

with $P_{\bar{H}} = \mathrm{diag}(B_{\mathrm{f}}, B_{\mathrm{c}})$. Further, using $\bar{P}_{\mathrm{fc}} := P_{\mathrm{fc}} B_{\mathrm{c}}$ the Schur complement S_{cc} of $P_{\bar{H}}^T \bar{H} P_{\bar{H}}$ is given by

$$\begin{aligned} S_{\mathrm{cc}} &= B_{\mathrm{c}}^T H_{\mathrm{cc}} B_{\mathrm{c}} + (1+c_0)\bar{P}_{\mathrm{fc}}^T H_{\mathrm{ff}} \bar{P}_{\mathrm{fc}} - \bar{P}_{\mathrm{fc}}^T H_{\mathrm{ff}} B_{\mathrm{f}} (B_{\mathrm{f}}^T H_{\mathrm{ff}} B_{\mathrm{f}})^{-1} B_{\mathrm{f}}^T H_{\mathrm{ff}} \bar{P}_{\mathrm{fc}} \\ &= B_{\mathrm{c}}^T H_{\mathrm{cc}} B_{\mathrm{c}} + c_0 \bar{P}_{\mathrm{fc}}^T H_{\mathrm{ff}} \bar{P}_{\mathrm{fc}} \end{aligned}$$

because $B_{\mathrm{c}}(B_{\mathrm{c}}^T B_{\mathrm{c}})^{-1} B_{\mathrm{c}}^T H_{\mathrm{ff}} = P_{\mathrm{f}} H_{\mathrm{ff}} = H_{\mathrm{ff}}$, since H_{ff} is symmetric. Thence, with $\bar{\bar{H}}_{\mathrm{cc}} := B_{\mathrm{c}}^T H_{\mathrm{cc}} B_{\mathrm{c}}$ we obtain

$$\begin{aligned} \gamma(\bar{H})^2 &= 1 - \min_{\boldsymbol{v}} \frac{\boldsymbol{v}^T S_{\mathrm{cc}} \boldsymbol{v}}{\boldsymbol{v}^T (P_{\bar{H}}^T \bar{H} P_{\bar{H}})_{\mathrm{cc}} \boldsymbol{v}} = 1 - \min_{\boldsymbol{v}} \frac{\boldsymbol{v}^T (\bar{\bar{H}}_{\mathrm{cc}} + c_0 \bar{P}_{\mathrm{fc}}^T H_{\mathrm{ff}} \bar{P}_{\mathrm{fc}}) \boldsymbol{v}}{\boldsymbol{v}^T (\bar{\bar{H}}_{\mathrm{cc}} + (1+c_0) \bar{P}_{\mathrm{fc}}^T H_{\mathrm{ff}} \bar{P}_{\mathrm{fc}}) \boldsymbol{v}} \\ &= 1 - \frac{1}{1 + \max_{\boldsymbol{v}} \frac{\boldsymbol{v}^T \bar{P}_{\mathrm{fc}}^T H_{\mathrm{ff}} \bar{P}_{\mathrm{fc}} \boldsymbol{v}}{\boldsymbol{v}^T (\bar{\bar{H}}_{\mathrm{cc}} + c_0 \bar{P}_{\mathrm{fc}}^T H_{\mathrm{ff}} \bar{P}_{\mathrm{fc}}) \boldsymbol{v}}} = 1 - \frac{1}{1 + \frac{1}{c_0 + \min_{\boldsymbol{v}} \frac{\boldsymbol{v}^T \bar{\bar{H}}_{\mathrm{cc}} \boldsymbol{v}}{\boldsymbol{v}^T \bar{P}_{\mathrm{fc}}^T H_{\mathrm{ff}} \bar{P}_{\mathrm{fc}} \boldsymbol{v}}}} \\ &= \frac{1}{1 + c_0 + \|\bar{\bar{H}}_{\mathrm{cc}}^{-1} \bar{P}_{\mathrm{fc}} H_{\mathrm{ff}} \bar{P}_{\mathrm{fc}}\|_2} \leq \frac{1}{1 + \|\bar{\bar{H}}_{\mathrm{cc}}^{-1} \bar{P}_{\mathrm{fc}} H_{\mathrm{ff}} \bar{P}_{\mathrm{fc}}\|_2}. \end{aligned}$$

The previous equations hold true if $H_{\mathrm{cc}} \neq 0$ and $H_{\mathrm{ff}} \neq 0$. For $H_{\mathrm{cc}} \equiv 0$ we conclude $\gamma(\bar{H})^2 = \frac{1}{1+c_0}$. On the other hand, if $H_{\mathrm{ff}} \equiv 0$ we have $\gamma(\bar{H}) = 0$ and hence $\gamma(\hat{B}) = 0$. Moreover, note that $\|\bar{\bar{H}}_{\mathrm{cc}}^{-1} \bar{P}_{\mathrm{fc}} H_{\mathrm{ff}} \bar{P}_{\mathrm{fc}}\|_2 = \lambda_{\max}(P_{\mathrm{fc}} H_{\mathrm{ff}} P_{\mathrm{fc}}, H_{\mathrm{cc}})$, which is the maximal generalized eigenvalue of $P_{\mathrm{fc}} H_{\mathrm{ff}} P_{\mathrm{fc}}$ with respect to H_{cc}.

Summarizing,

$$\gamma(\hat{B}) \leq \gamma(\bar{H}(c_0)) < 1. \tag{4.34}$$

For further discussions we need the following lemma, which states the spectral equivalence of the Schur complements of spectrally equivalent matrices.

Lemma 4.6.1. *Let A, B be $\mathbb{R}^{n \times n}$ matrices, $n \in \mathbb{N}$, partitioned into 2×2 blocks, see* (3.35). *Further, let A_{ff} and B_{ff} be regular and let A and B be spectrally equivalent, i.e.,*

$$\underline{c}(A\boldsymbol{v}, \boldsymbol{v}) \leq (B\boldsymbol{v}, \boldsymbol{v}) \leq \overline{c}(A\boldsymbol{v}, \boldsymbol{v}) \qquad \forall \boldsymbol{v} \in \mathbb{R}^n \tag{4.35}$$

for some $\overline{c} > \underline{c} > 0$. Then the Schur complement S_{cc}^B is spectrally equivalent to S_{cc}^A with the same constants $\underline{c}$ and $\overline{c}$.

Proof. The bounds follow from the minimization-property of the Schur complement and the spectral equivalence of A and B. For the upper bound we get

$$\boldsymbol{v}_c^T S_{cc}^B \boldsymbol{v}_c = \min_{\boldsymbol{v}_f} \begin{bmatrix} \boldsymbol{v}_f \\ \boldsymbol{v}_c \end{bmatrix}^T B \begin{bmatrix} \boldsymbol{v}_f \\ \boldsymbol{v}_c \end{bmatrix} \leq \bar{c} \min_{\boldsymbol{v}_f} \begin{bmatrix} \boldsymbol{v}_f \\ \boldsymbol{v}_c \end{bmatrix}^T A \begin{bmatrix} \boldsymbol{v}_f \\ \boldsymbol{v}_c \end{bmatrix} = \bar{c} \boldsymbol{v}_c^T S_{cc}^A \boldsymbol{v}_c \,.$$

The lower bound is obtained in an analogous manner. □

In view of the previous observations we are able to derive the following upper bound for $\gamma(\hat{G})$.

Theorem 4.6.2. *Let $\hat{G} = J^T G J$ and $\hat{B} = J^T B J$ be defined by* (4.10) *and* (4.33) *using* (3.38). *Further, let* (4.32) *be satisfied. Then, we get the following bound for the CBS constant $\gamma(\hat{G})$:*

$$\gamma(\hat{G}) \leq \sqrt{1 - (1 - \gamma(\bar{H})^2)\frac{\underline{c}}{\bar{c}}}\,. \tag{4.36}$$

Proof. First, note that (4.32) implies the spectral equivalence of $\hat{G}$ and $\hat{B}$ with the same constants $\underline{c}$ and $\bar{c}$. Now, let $B_f : \mathbb{R}^{\dim(\ker(\hat{G}_{ff})^\perp)} \to \ker(\hat{G}_{ff})^\perp$ and $B_c : \mathbb{R}^{\dim(\ker(\hat{G}_{cc})^\perp)} \to \ker(\hat{G}_{cc})^\perp$ denote basis matrices, mapping onto the orthogonal complements of the kernels of $\hat{G}_{ff}$ and $\hat{G}_{cc}$, respectively. Then we conclude for $\gamma(\hat{G})$

$$\begin{aligned}
\gamma(\hat{G}) &= \sup_{\substack{\boldsymbol{v}_f \notin \ker(\hat{G}_{ff}) \\ \boldsymbol{v}_c \notin \ker(\hat{G}_{cc})}} \frac{\boldsymbol{v}_f^T \hat{G}_{fc} \boldsymbol{v}_c}{\sqrt{\boldsymbol{v}_f^T \hat{G}_{ff} \boldsymbol{v}_f \cdot \boldsymbol{v}_c^T \hat{G}_{cc} \boldsymbol{v}_c}} \\
&= \sup_{\tilde{\boldsymbol{v}}_f, \tilde{\boldsymbol{v}}_c \neq 0} \frac{\tilde{\boldsymbol{v}}_f^T B_f^T \hat{G}_{fc} B_c \tilde{\boldsymbol{v}}_c}{\sqrt{\tilde{\boldsymbol{v}}_f^T B_f^T \hat{G}_{ff} B_f \tilde{\boldsymbol{v}}_f \cdot \tilde{\boldsymbol{v}}_c^T B_c^T \hat{G}_{cc} B_c \tilde{\boldsymbol{v}}_c}} = \gamma(\hat{P}^T \hat{G} \hat{P}) \,,
\end{aligned}$$

with $\hat{P} := \operatorname{diag}(B_f, B_c)$. Analogously, we get $\gamma(\hat{B}) = \gamma(\hat{P}^T \hat{B} \hat{P})$. Now, we obtain with Lemma 4.6.1 (using also Lemma 9.2 in [Axe94])

$$\begin{aligned}
1 - \gamma(\hat{G})^2 &= 1 - \gamma(\hat{P}^T \hat{G} \hat{P})^2 = \inf_{\boldsymbol{v}_c \neq 0} \frac{\boldsymbol{v}_c^T S_{cc}^{\hat{P}^T \hat{G} \hat{P}} \boldsymbol{v}_c}{\boldsymbol{v}_c P_c^T \hat{G}_{cc} P_c \boldsymbol{v}_c} \\
&\geq \frac{\underline{c}}{\bar{c}} \inf_{\boldsymbol{v}_c \neq 0} \frac{\boldsymbol{v}_c^T S_{cc}^{\hat{P}^T \hat{B} \hat{P}} \boldsymbol{v}_c}{\boldsymbol{v}_c P_c^T \hat{B}_{cc} P_c \boldsymbol{v}_c} = \frac{\underline{c}}{\bar{c}}(1 - \gamma(\hat{P}^T \hat{B} \hat{P})^2) = \frac{\underline{c}}{\bar{c}}(1 - \gamma(\hat{B})^2) \\
&\geq (1 - \gamma(\bar{H})^2)\frac{\underline{c}}{\bar{c}} \,,
\end{aligned}$$

from which (4.36) follows. □

Remark. So far we have investigated the CBS constant of $\hat{G} = J^T G J$, where G is defined by (4.10). From $\hat{A} = \hat{G} + \hat{F}$ we conclude that

$$\gamma(\hat{A}) \leq \max\{\gamma(\hat{G}),\, \gamma(\hat{F})\}\,.$$

Hence, assuming that $\gamma(\hat{F}) \leq \gamma(\hat{G})$ we obtain with (4.36)

$$\gamma(\hat{A}) \leq \sqrt{1 - (1 - \gamma(\bar{H})^2)\frac{\underline{c}}{\overline{c}}}\,. \tag{4.37}$$

In the remainder of this section we study by means of an example the spectral equivalence of A_T and B_T, i.e., the evaluation of $\gamma(\hat{G})$. We know from Theorem 4.2.2 that the edge matrices E_{ij} are of the form (4.9), that is, only c_{ij} is left to be determined. Therefore, we investigate the minimal relative condition number $\kappa(A_T,\, B_T)$ for the reference element, see Figure 4.3. In the two-dimensional case we compare the best possible $\kappa(A_T,\, B_T)$,[4] depicted by the solid line, to the relative condition number that is obtained when choosing c_{ij} according to (4.16). The latter curve is dotted. We see that using (4.16) yields a condition number which is quite close to the optimum. Additionally to the condition numbers in 2D, the dashed line in Figure 4.3 describes $\kappa(A_T,\, B_T)$ obtained for the reference tetrahedron in 3D. For reasonable ν we still get acceptable results. Furthermore we observe the large slope of κ when ν approaches 1/2. This is due to the ill-posedness of the system (4.2) in the limit case $\nu = 1/2$. For instance, we have $\kappa(A_T,\, B_T) = 11.2$ (22.6) for $\nu = 0.45$ in 2D (3D).

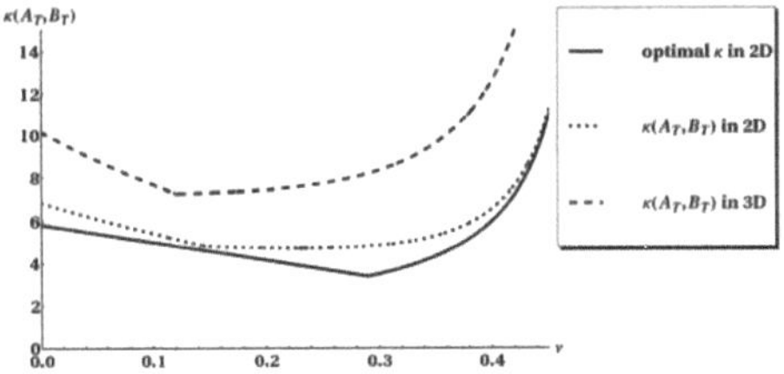

Figure 4.3: The general condition numbers $\kappa(A_T,\, B_T)$ for optimal constants c_{ij} (solid line) and for the constants according to AMGm in 2D (dotted line) and 3D (dashed line).

4.7 Aspects on parallelizing AMGm

In the following we briefly address some implementation issues related to a parallelization of the proposed method.

[4] The best possible κ was computed by minimizing $\kappa(A_T,\, B_T)$ with respect to the c_{ij}.

1. The mesh has to be partitioned into a set of submeshes using a mesh-partitioning software. Further we need an overlap of the distributed mesh. The overlap of submeshes has to contain all elements that share an edge with the boundary of any subdomain.

2. After setting up the matrices $A_{h,p}$, $G_{h,p}$ and $F_{h,p}$, the edge matrices on each processor p can be computed independently. Note that the edge matrices of any edge on the subdomain boundary coincide on each of the corresponding processors due to the overlap. The edge matrices of the elements belonging to the overlap have to be distributed. Thereby, only the single values c_{ij} need to be broadcasted for each edge.

3. The computation of the strength of connectivity s_{ij} for each edge e_{ij} of a subdomain can be preformed without communication, due to the knowledge of the edge matrices of the overlap.

4. The partition into strong and weak edges on each processor is quite simple. Only the choice of strong and weak edges on the subdomain boundary needs to be synchronized. The reason for this are the different threshold values θ_p on different subdomains since the ratio Θ of weak and strong edges is computed separately on each subdomain or processor. A similar problem occurs in the coarse-grid selection step where one can define a vertex on the subdomain boundary to be coarse if it has been select as a coarse vertex on at least one subdomain.

5. The interpolation of fine interior points is as explained in Section 4.5. For fine boundary vertices we can use solely strong fine-coarse connections for interpolation. This guarantees a uniquely defined global interpolation.

6. After building the coarse matrices $A_{H,p}$ by the Galerkin product the coarse edges have to be determined. First each processor determines the interior coarse edges. At a second stage those coarse edges for which at least one vertex is in the overlap of two subdomains are computed. They are distributed to the corresponding processors. There will never be an edge connecting an interior vertex of one subdomain with an interior vertex of another subdomain!

The smoother in a parallel version could be a symmetric Gauss-Seidel iteration within each subdomain completed by a global block Jacobi-type smoothing, which is a standard option in BoomerAMG (see [HMY00]).

4.8 Numerical results

In the final section we discuss the numerical results obtained by applying the proposed (modified) AMGm method to selected linear elasticity problems. Basically, we consider three types of test problems. The first one is a composite material consisting of two different matters which differ in their Young's modulus. The second example is a three-dimensional beam. On this example we also present a comparison of methods including the original AMGm method (as proposed in [Kra08]) and the BoomerAMG ([HMY00]). In the last part of this section we are dealing with orthotropic materials, such as wood and cancellous bones.

AMGm has been implemented as a preconditioner for the preconditioned conjugate gradient (PCG) method in the finite element software package NGSolve (see [Sch97]). As a smoother we use a symmetric block Gauss-Seidel method. We perform tests with the V- and with the W-cycle AMGm preconditioner with one pre- and one post-smoothing step, denoted by V(1,1) and W(1,1). The convergence properties of the solver are presented in tabular form by listing the average convergence factor (ρ) as well as the number of iterations (#it) that are required to reduce the initial residual in the norm $\|\cdot\|_{B_{\mathrm{AMG}}^{-1}}$ by a factor 10^8. The numbers in parenthesis refer to edge matrices that are computed directly from A instead of G whereas the numbers without parenthesis are the results obtained when using the auxiliary matrix G for the computation of the edge matrices, as used in the two-level convergence analysis (in Section 4.6). Further, we report the grid and operator complexities σ^{Ω} and σ^{A}.[5] All the computations are performed on three different meshes; The two finer meshes have been generated from the coarsest mesh by uniform refinement. Finally, we mention that we use a fraction $\Theta = 0.08$ of weak edges.

4.8.1 Composite material

In this example we consider a composite material, which is distributed in the unit cube $\Omega = (0, 1)^3$ according to a checkerboard pattern, as depicted in Figure 4.4(a). We solve Problem 4.1.1 where the same Poisson ratio ν, but different moduli of elasticity E, referred to as E_0 and E_1, are used. Note that only the ratio E_1/E_0 is of importance in this example.

We fix the displacement $\boldsymbol{u} = 0$ at the bottom face $\Gamma_D := [0, 1] \times [0, 1] \times \{0\}$. Further, we apply a force in the negative z-direction on the top face $\Gamma_{N,1} := [0, 1] \times [0, 1] \times \{1\}$ and

[5] σ^{Ω} denotes the ratio of the total number of vertices (nodes) on all levels and the number of vertices on the finest grid; σ^{A}, on the other hand, is the total number of nonzeros in the matrices (on all levels) divided by the number of nonzeros in the fine-grid matrix.

impose homogeneous Neumann boundary conditions on the remaining part of the boundary $\Gamma_{N,0} := \partial\Omega \backslash (\Gamma_D \cup \Gamma_{N,1})$, which results in

$$\boldsymbol{t}_N = \begin{cases} \boldsymbol{0} & \text{on } \Gamma_{N,0} \\ (0,\, 0,\, -t_z)^T & \text{on } \Gamma_{N,1} \end{cases} . \tag{4.38}$$

The computations are performed for varying ratio E_1/E_0, ranging from 1 to 1 000, on three

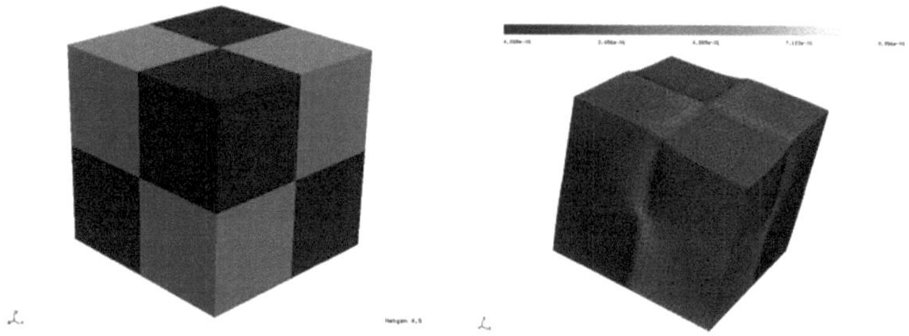

(a) geometry (b) solution for $E_1/E_0 = 10$ and $\nu = 0.2$

Figure 4.4: Geometry and solution for the composed cube.

different meshes. Additionally, we choose two values for the Poisson ratio ν, namely $\nu = 0.2$ and $\nu = 0.4$. In Figure 4.4(b) the solution for $\nu = 0.2$ and $E_1/E_0 = 10$ is illustrated.

In Table 4.1 and Table 4.2 the convergence properties and complexities of the method are reported for $\nu = 0.2$ and $\nu = 0.4$, respectively. At first we observe that the W-cycle AMGm preconditioner results in convergence rates that are independent of the mesh size h. The operator complexity σ^A is almost constant. For the V-cycle method, we observe a dependence of the iteration count on the mesh size h, i.e., on the number of levels n_L. Under weak assumptions the general condition number for the V-cycle preconditioner typically depends on the number of levels n_L, see, e.g., [Vas08].[6] For a detailed discussion we refer to [Vas08]. The second important result is the robustness with respect to the ratio E_1/E_0. Throughout both tables we can clearly see that this quotient has almost no influence on the convergence of the method. Thirdly, when comparing Table 4.1 to Table 4.2 we observe a moderate increase of the number of iterations for $\nu = 0.4$. This is due to the deterioration of $\kappa(A_T, B_T)$ defined by (4.7), which can be seen in Figure 4.3.

[6]Only under a stronger "smoothing property" an h- and n_L-independent convergence rate can be proven.

#elements (#DOF)		22 043 (14 028)		176 344 (99 639)		1 410 752 (750 135)	
#levels		4		6		8	
$\nu = 0.2$		#it.	ρ	#it.	ρ	#it.	ρ
$E_1/E_0 = 1$:	V(1,1)	38 (28)	0.61 (0.51)	56 (42)	0.72 (0.64)	68 (50)	0.76 (0.69)
	W(1,1)	18 (13)	0.33 (0.24)	21 (13)	0.40 (0.24)	19 (14)	0.36 (0.24)
	σ^Ω	1.56 (1.59)		1.50 (1.50)		1.48 (1.48)	
	σ^A	3.38 (3.49)		3.41 (3.45)		3.58 (3.56)	
$E_1/E_0 = 10$:	V(1,1)	44 (39)	0.65 (0.62)	66 (59)	0.76 (0.73)	88 (60)	0.81 (0.73)
	W(1,1)	20 (17)	0.37 (0.31)	20 (17)	0.40 (0.31)	18 (14)	0.34 (0.25)
	σ^Ω	1.57 (1.56)		1.50 (1.50)		1.48 (1.48)	
	σ^A	3.41 (3.31)		3.38 (3.44)		3.58 (3.57)	
$E_1/E_0 = 100$:	V(1,1)	51 (39)	0.70 (0.62)	71 (70)	0.77 (0.76)	77 (73)	0.79 (0.78)
	W(1,1)	19 (15)	0.38 (0.27)	19 (17)	0.36 (0.33)	17 (14)	0.32 (0.25)
	σ^Ω	1.56 (1.56)		1.50 (1.50)		1.48 (1.48)	
	σ^A	3.33 (3.34)		3.44 (3.43)		3.58 (3.57)	
$E_1/E_0 = 1\,000$:	V(1,1)	42 (36)	0.64 (0.60)	78 (75)	0.79 (0.78)	79 (84)	0.79 (0.80)
	W(1,1)	16 (14)	0.31 (0.25)	19 (18)	0.36 (0.35)	15 (15)	0.28 (0.27)
	σ^Ω	1.58 (1.57)		1.50 (1.50)		1.48 (1.48)	
	σ^A	3.45 (3.38)		3.40 (3.43)		3.58 (3.58)	

Table 4.1: Comparison of $V(1,1)$ and $W(1,1)$-cycle AMGm-PCG for $\varepsilon = 10^{-8}$ and $\nu = 0.2$: Composite material.

#elements (#DOF)		22 043 (14 028)		176 344 (99 639)		1 410 752 (750 135)	
#levels		4		6		8	
$\nu = 0.4$		#it.	ρ	#it.	ρ	#it.	ρ
$E_1/E_0 = 1$:	V(1,1)	46 (33)	0.67 (0.57)	63 (53)	0.74 (0.70)	73 (61)	0.78 (0.74)
	W(1,1)	22 (16)	0.42 (0.31)	23 (17)	0.44 (0.33)	22 (16)	0.42 (0.31)
	σ^Ω	1.56 (1.58)		1.51 (1.51)		1.48 (1.48)	
	σ^A	3.35 (3.45)		3.45 (3.43)		3.49 (3.49)	
$E_1/E_0 = 10$:	V(1,1)	52 (44)	0.70 (0.65)	83 (56)	0.80 (0.71)	78 (68)	0.79 (0.76)
	W(1,1)	23 (20)	0.45 (0.38)	25 (16)	0.47 (0.30)	20 (17)	0.39 (0.32)
	σ^Ω	1.56 (1.57)		1.50 (1.50)		1.48 (1.48)	
	σ^A	3.34 (3.44)		3.42 (3.41)		3.51 (3.49)	
$E_1/E_0 = 100$:	V(1,1)	58 (50)	0.73 (0.69)	81 (68)	0.79 (0.76)	97 (86)	0.83 (0.81)
	W(1,1)	23 (20)	0.43 (0.39)	22 (17)	0.42 (0.34)	19 (16)	0.36 (0.31)
	σ^Ω	1.56 (1.57)		1.51 (1.51)		1.48 (1.48)	
	σ^A	3.37 (3.41)		3.45 (3.44)		3.50 (3.49)	
$E_1/E_0 = 1000$:	V(1,1)	45 (40)	0.66 (0.63)	84 (55)	0.80 (0.71)	110 (97)	0.84 (0.83)
	W(1,1)	18 (15)	0.35 (0.29)	21 (15)	0.40 (0.28)	19 (17)	0.38 (0.30)
	σ^Ω	1.56 (1.57)		1.51 (1.51)		1.48 (1.48)	
	σ^A	3.34 (3.41)		3.47 (3.47)		3.51 (3.49)	

Table 4.2: Comparison of $V(1,1)$ and $W(1,1)$-cycle of PCG for $\varepsilon = 10^{-8}$ and $\nu = 0.4$ applied to a composite material.

4.8.2 3D beam

Now, we investigate the influence of the shape of the reference configuration by considering a beam, i.e., $\Omega = (0, 20) \times (0, 2) \times (0, 1)$ consisting of an isotropic material ($C = C_{\text{iso}}$ defined by (2.47)). We fix the deformation on both ends of the beam to be zero, that is,

$$\boldsymbol{u} = \boldsymbol{0} \qquad \text{on } \Gamma_D := \{0, 20\} \times [0, 2] \times [0, 1] .$$

Moreover, we apply a force in negative z-direction on a circle $\Gamma_{N,1} := \{\boldsymbol{x} : (x_1 - 10)^2 + (x_2 - 1)^2 \leq 0.25 \wedge x_3 = 1\}$. On the remaining part of the boundary $\Gamma_{N,0} := \partial\Omega \backslash \{\Gamma_D \cup \Gamma_{N,1}\}$ we impose homogenous Neumann boundary conditions. Hence $\boldsymbol{t}_N$ is given by (4.38) for some $t_z > 0$. In Figure 4.5 the solution of the system (4.2) is shown (for the smallest problem size). The shading indicates the absolute value of the stress.

In order to demonstrate the improvement of the method from [Kra08] by the modifications that have been presented in this chapter,i.e., in [KK10], the results of the original AMGm version are listed for comparison. Further, to have a comparison with another state-of-the-art method, the results obtained with BoomerAMG (see [HMY00]) are included.

We report the convergence history of the three methods in Table 4.3 for $\nu = 0.1, 0.25, 0.4$. The considered examples contain up to 6 million DOF on the finest level in order to show scalability of AMG in terms of the problem size. The parameters of the old AMGm version have been chosen according to [Kra08] (example reported in Table 5.7). BoomerAMG was used in the system's version with the PMIS coarsening (see [DSMYH06]), which typically results in a low operator complexity.

One can clearly see the improvement of the (new) AMGm method in terms of convergence rates. The operator complexity is also slightly reduced as compared to the original (old) version. When we compare AMGm to BoomerAMG we find that while the W-cycle of AMGm is stable the iteration count of BoomerAMG's W-cycle slightly increases with respect to the mesh size and that the numbers of iterations which are needed are significantly higher than those of AMGm. This might be due to the much lower operator complexity of BoomerAMG.

In Table 4.4 we depict the time needed for the single methods to set up its preconditioner and additionally, how long it took to converge for the two considered cycles. We see that BoomerAMG sets up much faster, but due to the increased number of iterations it needs much longer to solve the problem. In total, BoomerAMG is still faster than AMGm. Nevertheless, for the considered example BoomerAMG scales worse than our method as can be seen in Table 4.5. Thereby, the ratios of set up and solution time between the three different meshes are shown. Especially the set up of our procedure shows a very good performance with respect to this measure.

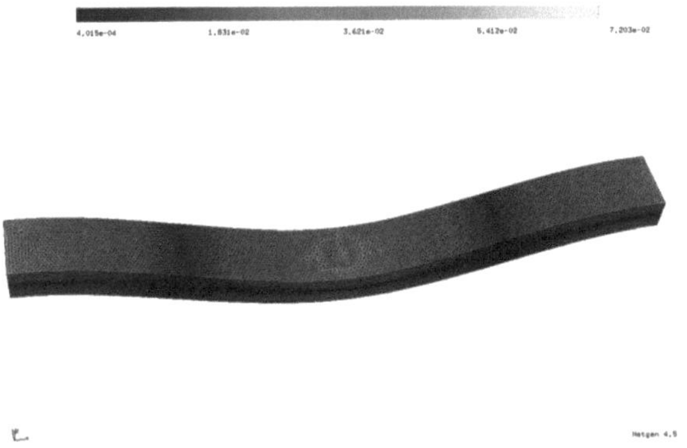

Figure 4.5: The deformed beam.

4.8.3 Orthotropic materials

In our last example we treat orthotropic materials with $C = C_{\text{ortho}}$ defined by (2.48). We consider system (4.2) for three different materials, which are cancellous bone, hard and soft wood. The parameter settings for these materials are taken from [YKvR+99] and for the cancellous bone additionally from [KvRD+99]. As in Subsection 4.8.1 we consider as the reference configuration the unit cube $\Omega := (0,\, 1)^3$, which is fixed on the bottom face $\Gamma_D := [0,\, 1]^2 \times \{0\}$ and to which a force is applied on the top face $\Gamma_{N,1} := [0,\, 1]^2 \times \{1\}$. Moreover, we have homogeneous Neumann boundary conditions on the remaining part of the boundary $\Gamma_{N,0} := \partial\Omega \backslash (\Gamma_D \cup \Gamma_{N,1})$, which yields

$$\boldsymbol{t}_N = \begin{cases} \mathbf{0} & \text{on } \Gamma_{N,0} \\ t_N \cdot (1,\, 1,\, -1)^T & \text{on } \Gamma_{N,1} \end{cases},$$

with $t_N > 0$. In Table 4.6 we list the parameters for C_{ortho} as given in [YKvR+99]. Thereby ϕ denotes the volume fraction of the cancellous bone, that is, the ratio between the volume of the solid phase of the bone and its total volume. Additionally, we identify E_t with the isotropic tissue modulus. We choose $E_t = 5.4\,\text{GPa}$ throughout all computations, which is an average value of different types of bones (cf. [KvRD+99]). According to [YKvR+99], the apparent density ρ of wood is related to ϕ via $\rho = \gamma\,\phi$ with $\gamma \approx 1.9\,\text{g/cm}^3$.

# elements (#DOF)		175 637 (109 533)		1 405 096 (785 676)		11 240 768 (5 945 445)	
$\nu = 0.10$		#it.	ρ	#it.	ρ	#it.	ρ
AMGm:	V(1,1)	68	0.76	121	0.86	136	0.87
	W(1,1)	17	0.32	18	0.34	15	0.27
	σ^A (σ^Ω)	3.33 (1.55)		3.34 (1.49)		3.31 (1.46)	
AMGm ([Kra08]):	V(1,1)	118	0.86	216	0.92	-	-
	W(1,1)	32	0.56	33	0.57	-	-
	σ^A (σ^Ω)	3.50 (1.55)		3.54 (1.50)		-	
BoomerAMG:	V(1,1)	59	0.73	69	0.76	84	0.80
	W(1,1)	45	0.66	48	0.68	54	0.71
	σ^A (σ^Ω)	1.91 (1.30)		1.88 (1.27)		2.08 (1.30)	
$\nu = 0.25$		#it.	ρ	#it.	ρ	#it.	ρ
AMGm:	V(1,1)	61	0.74	107	0.84	117	0.85
	W(1,1)	15	0.29	17	0.33	16	0.29
	σ^A (σ^Ω)	3.29 (1.54)		3.32 (1.48)		3.27 (1.46)	
AMGm ([Kra08]):	V(1,1)	122	0.86	233	0.92	-	-
	W(1,1)	33	0.57	35	0.59	-	-
	σ^A (σ^Ω)	3.50 (1.55)		3.54 (1.50)		-	
BoomerAMG:	V(1,1)	57	0.72	70	0.77	82	0.80
	W(1,1)	49	0.68	52	0.70	56	0.72
	σ^A (σ^Ω)	1.96 (1.32)		1.95 (1.28)		2.15 (1.32)	
$\nu = 0.40$		#it.	ρ	#it.	ρ	#it.	ρ
AMGm:	V(1,1)	70	0.77	124	0.86	146	0.88
	W(1,1)	18	0.35	18	0.35	18	0.35
	σ^A (σ^Ω)	3.29 (1.54)		3.29 (1.49)		3.25 (1.46)	
AMGm ([Kra08]):	V(1,1)	145	0.88	133	0.87	-	-
	W(1,1)	40	0.63	43	0.65	-	-
	σ^A (σ^Ω)	3.51 (1.55)		3.51 (1.49)		-	
BoomerAMG:	V(1,1)	61	0.74	74	0.78	90	0.81
	W(1,1)	51	0.69	52	0.70	62	0.74
	σ^A (σ^Ω)	2.13 (1.35)		2.13 (1.32)		2.32 (1.35)	

Table 4.3: Comparison of AMGm (old and new) and BoomerAMG; $V(1,1)$- and $W(1,1)$-cycle PCG for $\varepsilon = 10^{-8}$: 3D beam.

In the following we examine the influence of the degree of anisotropy ω, which we define as the maximum ratio of the Young's moduli, that is, $\omega := \max_{i \neq j \in \{1,2,3\}} E_i/E_j$. In Figure 4.6(a) ω is depicted for varying volume fraction ϕ. For soft wood we find an increase of the anisotropy for decreasing ϕ. For $\phi \to 0$ the degree of anisotropy ω converges to infinity for all materials due to different exponents of ϕ and ρ in the expressions for E_i, see Table 4.6. For comparison, in Figure 4.6(b) the relative condition number $\kappa(A_T, B_T)$ on the reference element is plotted against ϕ. Note that the experimental data, used in [YKvR+99] to derive the model (see Table 4.6), was available only for $\phi \in (0.05, 0.40)$.

For the numerical tests, we use two values of ϕ, i.e., $\phi = 0.3$ and $\phi = 0.1$. The corresponding values of C_{ortho} are listed in Table 4.7. The solution of Problem 4.1.1 for $\phi = 0.1$ on a mesh of approximately 33 000 vertices is shown in Figure 4.7 for soft and hard wood, as well as in Figure 4.8 for cancellous bone. We choose $t_N = 2\,\text{kN}$ for a cube with a side length of 1 cm.

# elements (#DOF)		175 637 (109 533)		1 405 096 (785 676)		11 240 768 (5 945 445)	
$\nu = 0.10$		setup	solve	setup	solve	setup	solve
AMGm:	V(1,1)	44.0	13.1	387	243	3 142	3 003
	W(1,1)		15.7		176		1 671
AMGm ([Kra08]):	V(1,1)	24.5	72.6	283	937	-	-
	W(1,1)		47.1		493		-
BoomerAMG:	V(1,1)	3.3	21.4	37.3	254	442	3 363
	W(1,1)		21.1		226		2 993
$\nu = 0.25$							
AMGm:	V(1,1)	43.2	11.6	373	214	3 032	2 583
	W(1,1)		13.7		167		1 758
AMGm ([Kra08]):	V(1,1)	25.1	65.8	281	1 062	-	-
	W(1,1)		61.9		564		-
BoomerAMG:	V(1,1)	3.3	21.3	38.5	268	450	3 407
	W(1,1)		22.2		253		3 268
$\nu = 0.40$							
AMGm:	V(1,1)	43.1	13.4	364	245	2 988	3 164
	W(1,1)		16.1		172		1 949
AMGm ([Kra08]):	V(1,1)	24.7	69.7	270	574	-	-
	W(1,1)		59.3		685		-
BoomerAMG:	V(1,1)	3.4	23.1	41.1	304	481	4 190
	W(1,1)		25.5		283		4 206

Table 4.4: Timing of AMGm (old and new) and BoomerAMG in seconds; $V(1,1)$- and $W(1,1)$-cycle PCG for $\varepsilon = 10^{-8}$: 3D beam.

ratio of #DOF of consecutive meshes			7.17		7.57	
			setup	solve	setup	solve
$\nu = 0.10$	AMGm:	V(1,1)	8.8	18.5	8.11	12.4
		W(1,1)		11.2		9.5
	AMGm ([Kra08]):	V(1,1)	11.6	12.9	-	-
		W(1,1)		10.5		-
	BoomerAMG:	V(1,1)	11.3	11.9	11.8	13.2
		W(1,1)		10.7		13.2
$\nu = 0.25$	AMGm:	V(1,1)	8.6	18.4	8.13	12.1
		W(1,1)		12.2		10.5
	AMGm ([Kra08]):	V(1,1)	11.2	16.1	-	-
		W(1,1)		9.1		-
	BoomerAMG:	V(1,1)	11.7	12.6	11.7	12.7
		W(1,1)		11.4		12.9
$\nu = 0.40$	AMGm:	V(1,1)	8.4	18.3	8.21	12.9
		W(1,1)		10.7		11.3
	AMGm ([Kra08]):	V(1,1)	10.9	8.2	-	-
		W(1,1)		11.6		-
	BoomerAMG:	V(1,1)	12.1	13.2	11.7	13.8
		W(1,1)		11.1		14.9

Table 4.5: Scaling of AMGm (old and new) and BoomerAMG with respect to consecutive meshes; $V(1,1)$- and $W(1,1)$-cycle PCG for $\varepsilon = 10^{-8}$: 3D beam.

	cancellous bone	hard wood	soft wood
E_1	$1\,240\,E_t\,\phi^{1.80}$	$1.307\,\rho^{0.89}$ GPa	$2.05\,\rho^{1.71}$ GPa
E_2	$885\,E_t\,\phi^{1.89}$	$2.97\,\rho^{1.50}$ GPa	$3.14\,\rho^{1.59}$ GPa
E_3	$528.8\,E_t\,\phi^{1.92}$	$27.63\,\rho^{1.41}$ GPa	$32.01\,\rho^{1.01}$ GPa
μ_{12}	$486.3\,E_t\,\phi^{1.98}$	$0.4125\,\rho^{1.21}$ GPa	$0.083\,\rho^{0.66}$ GPa
μ_{13}	$316.65\,E_t\,\phi^{1.97}$	$1.57\,\rho^{1.37}$ GPa	$2.05\,\rho^{1.36}$ GPa
μ_{23}	$266.65\,E_t\,\phi^{2.04}$	$1.97\,\rho^{1.23}$ GPa	$2.28\,\rho^{1.27}$ GPa
ν_{12}	$0.176\,E_t\,\phi^{-0.25}$	$0.724\,\rho^{0.90}$	$0.269\,\rho^{-0.17}$
ν_{13}	$0.316\,E_t\,\phi^{-0.19}$	$0.016\,\rho^{-0.76}$	$0.019\,\rho^{0.10}$
ν_{23}	$0.256\,E_t\,\phi^{-0.09}$	$0.024\,\rho^{-0.73}$	$0.028\,\rho^{0.18}$

Table 4.6: Parameters for the stress-strain matrix C_{ortho} for three different materials.

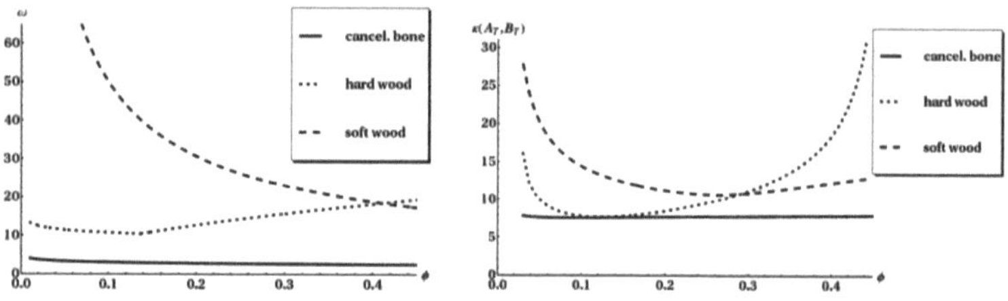

(a) the anisotropy ω (b) the relative condition number $\kappa(A_T, B_T)$

Figure 4.6: Anisotropy and the relative condition number for the 3 materials and varying ϕ.

One can clearly see a stronger deformation of the hard wood, while the cancellous bone does almost not deform due to the higher Young's moduli E_i.

	$\phi = 0.3$			$\phi = 0.1$		
	cancel. bone	hard wood	soft wood	cancel. bone	hard wood	soft wood
E_1	766.7	0.793	0.784	106.1	0.298	0.120
E_2	491.0	1.278	1.285	61.57	0.246	0.224
E_3	283.0	12.51	18.14	34.33	2.657	5.982
μ_{12}	242.1	0.209	0.057	27.50	0.055	0.028
μ_{13}	159.6	0.727	0.954	18.32	0.161	0.214
μ_{23}	123.5	0.987	1.117	13.13	0.255	0.277
ν_{12}	0.238	0.437	0.296	0.313	0.162	0.357
ν_{13}	0.397	0.025	0.018	0.489	0.057	0.016
ν_{23}	0.285	0.036	0.025	0.315	0.081	0.021

Table 4.7: Parameters for the stress-strain matrix C_{ortho} for three different materials. The moduli E_i and μ_{ij} are given in GPa. $E_t = 5.4\,\text{GPa}$ and $\rho = \gamma\,\phi$ with $\gamma = 1.9\text{g/cm}^3$ (constant).

Finally, in Table 4.8 and Table 4.9 the computational results are listed. First we note the stable operator complexity σ^A. Furthermore we observe uniform convergence with respect to

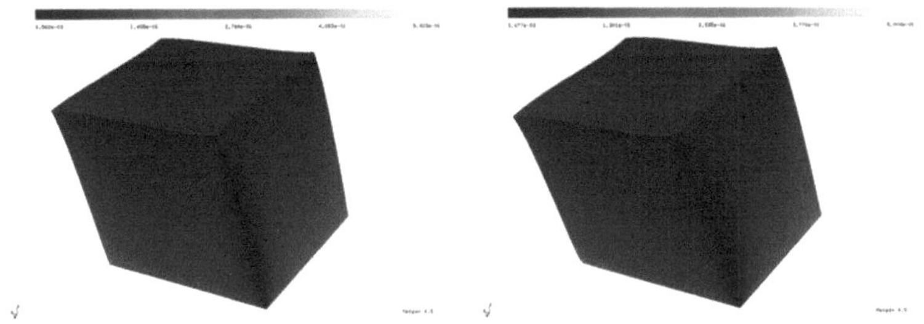

(a) soft wood (b) hard wood

Figure 4.7: The deformed cube using the stress-strain matrix of two different orthotropic materials, i.e., two types of wood. The shading corresponds to the absolute value of the stress.

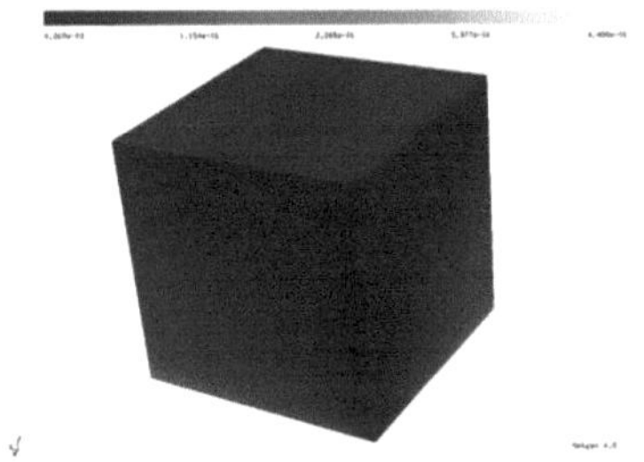

Figure 4.8: The deformed cube for cancellous bone.

the mesh size h and the number of levels n_L for all considered materials when employing the $W(1,1)$-cycle preconditioner. The convergence behavior is reflected in the relative condition numbers of the edge matrix approximations, see Figure 4.6(b). For the cancellous bone the number of iterations slightly increases when changing ϕ from 0.3 to 0.1. On the other hand, the convergence in the latter case is even faster for hard wood. The results for soft wood are similar for both settings.

#elements (#DOF) #levels		21 473 (13 992) 4		171 784 (98 259) 6		1 374 272 (735 303) 8	
		#it.	ρ	#it.	ρ	#it.	ρ
bone:	V(1,1)	34 (32)	0.58 (0.56)	49 (51)	0.68 (0.69)	57 (58)	0.72 (0.73)
	W(1,1)	16 (14)	0.30 (0.24)	16 (15)	0.30 (0.27)	14 (13)	0.26 (0.22)
	σ^{Ω}	1.58 (1.58)		1.51 (1.51)		1.48 (1.48)	
	σ^{A}	3.41 (3.49)		3.42 (3.42)		3.39 (3.39)	
hard wood:	V(1,1)	42 (35)	0.64 (0.59)	65 (61)	0.75 (0.74)	78 (67)	0.79 (0.76)
	W(1,1)	19 (15)	0.36 (0.29)	24 (20)	0.45 (0.40)	21 (18)	0.41 (0.34)
	σ^{Ω}	1.59 (1.58)		1.51 (1.51)		1.48 (1.49)	
	σ^{A}	3.54 (3.42)		3.46 (3.44)		3.47 (3.47)	
soft wood:	V(1,1)	51 (44)	0.69 (0.65)	81 (75)	0.79 (0.78)	100 (86)	0.83 (0.81)
	W(1,1)	24 (22)	0.46 (0.41)	30 (29)	0.53 (0.52)	29 (27)	0.53 (0.50)
	σ^{Ω}	1.60 (1.59)		1.52 (1.52)		1.49 (1.49)	
	σ^{A}	3.57 (3.51)		3.58 (3.54)		3.57 (3.58)	

Table 4.8: Comparison of $V(1,1)$- and $W(1,1)$-cycle AMGm-PCG for $\varepsilon = 10^{-8}$: Different orthotropic materials and a volume fraction $\phi = 0.3$.

#elements (#DOF) #levels		21 473 (13 992) 4		171 784 (98 259) 6		1 374 272 (735 303) 8	
		#it.	ρ	#it.	ρ	#it.	ρ
bone:	V(1,1)	37 (34)	0.61 (0.57)	58 (51)	0.73 (0.70)	71 (61)	0.77 (0.74)
	W(1,1)	17 (15)	0.34 (0.28)	20 (16)	0.37 (0.32)	19 (15)	0.36 (0.28)
	σ^{Ω}	1.59 (1.58)		1.51 (1.51)		1.48 (1.48)	
	σ^{A}	3.59 (3.41)		3.44 (3.41)		3.37 (3.38)	
hard wood:	V(1,1)	41 (34)	0.63 (0.58)	63 (54)	0.74 (0.71)	79 (63)	0.79 (0.74)
	W(1,1)	18 (16)	0.36 (0.30)	23 (18)	0.45 (0.36)	22 (17)	0.43 (0.33)
	σ^{Ω}	1.58 (1.59)		1.51 (1.52)		1.49 (1.49)	
	σ^{A}	3.39 (3.48)		3.51 (3.53)		3.53 (3.53)	
soft wood:	V(1,1)	46 (44)	0.67 (0.66)	83 (76)	0.80 (0.78)	103 (80)	0.84 (0.79)
	W(1,1)	21 (21)	0.40 (0.40)	30 (26)	0.53 (0.49)	26 (25)	0.49 (0.47)
	σ^{Ω}	1.59 (1.59)		1.52 (1.52)		1.50 (1.50)	
	σ^{A}	3.46 (3.49)		3.60 (3.57)		3.63 (3.61)	

Table 4.9: Comparison of $V(1,1)$- and $W(1,1)$-cycle AMGm-PCG for $\varepsilon = 10^{-8}$: Different orthotropic materials and a volume fraction $\phi = 0.1$.

4.9 Application to DG discretizations

Now, let us investigate the problem discussed in Subsection 2.4.3. Therefore, we consider problem (2.24) with $V = \boldsymbol{V}_h^{DG}$ and $a(.,.)$ given by (2.57).

4.9.1 Setup

For any face $E \in \mathcal{E}_h$ we denote by $T_{E,1} \in \mathcal{T}_h$ and $T_{E,2} \in \mathcal{T}_h$ its adjacent elements. If $E \in \mathcal{E}_h^D \cup \mathcal{E}_h^N$ we set $T_{E,2} = \emptyset$. From (2.57) it easily follows that $a(.,.)$ can be rewritten as

$$a(\boldsymbol{u}_{DG}, \boldsymbol{v}_{DG}) = \sum_{E \in \mathcal{E}_h^o \cup \mathcal{E}_h^D} a_E(\boldsymbol{u}_{DG}, \boldsymbol{v}_{DG}) \qquad \forall \boldsymbol{u}_{DG}, \boldsymbol{v}_{DG} \in \boldsymbol{V}_h^{DG},$$

with

$$\begin{aligned} a_E(\boldsymbol{u}_{DG}, \boldsymbol{v}_{DG}) \;=\; & \frac{1}{\delta_{T_{E,1}}}(\boldsymbol{\sigma}(\boldsymbol{u}_{DG}), \boldsymbol{\varepsilon}(\boldsymbol{u}_{DG}))_{0,T_{E,1}} + \frac{1}{\delta_{T_{E,2}}}(\boldsymbol{\sigma}(\boldsymbol{u}_{DG}, \boldsymbol{\varepsilon}(\boldsymbol{u}_{DG}))_{0,T_{E,2}} \\ & - \Big[(\{\!\{\boldsymbol{\sigma}(\boldsymbol{u}_{DG}) \cdot \boldsymbol{n}_E\}\!\}, [\![\boldsymbol{v}_{DG}]\!])_{0,E} + (\{\!\{\boldsymbol{\sigma}(\boldsymbol{v}_{DG}) \cdot \boldsymbol{n}_E\}\!\}, [\![\boldsymbol{u}_{DG}]\!])_{0,E}\Big] \\ & +(2\mu + \lambda)\gamma_0 \left(h^{-1} [\![P_0 \boldsymbol{u}_{DG}]\!], [\![P_0 \boldsymbol{v}_{DG}]\!]\right)_{0,E} \\ & +2\mu\gamma_1 \left(h^{-1} [\![\boldsymbol{u}_{DG}]\!], [\![\boldsymbol{v}_{DG}]\!]\right)_{0,E}, \end{aligned}$$

where $E \in \mathcal{E}_h^o \cup \mathcal{E}_h^D$. The constant δ_T is defined by

$$\delta_T = |\mathcal{E}_T| - |\mathcal{E}_T \cap \mathcal{E}_h^N| = 1 + d - |\mathcal{E}_T \cap \mathcal{E}_h^N|,$$

i.e., the number of faces of T minus the number of edges of T belonging to the Neumann boundary Γ_N. This induces a splitting of the matrix A into SPSD "edge matrices"[7] A_E with

$$A = \sum_{E \in \mathcal{E}_h^o \cup \mathcal{E}_h^D} A_E .$$

Finally, we want to shift the matrices A_E, belonging to $E \in \mathcal{E}_h^D$, to edge matrices A_E corresponding to interior edges. That is, for every $E \in \mathcal{E}_h^D$ we adapt all $A_{E'}$ for $E' \in \mathcal{E}_h^o \cap \mathcal{E}_{T_{E,1}}$ according to

$$A_{E'} = A_{E'} + \frac{1}{|\mathcal{E}_h^o \cap \mathcal{E}_{T_{E,1}}|} A_E . \tag{4.39}$$

Those matrices turn out to be SPD. We arrive at the decomposition of A through $A = \sum_{E \in \mathcal{E}_h^o} A_E$. Now, let us revisit the setting of Section 4.2. Let us define the set $\mathcal{V}$ to contain n_T algebraic vertices $\boldsymbol{v}_i$. Each $\boldsymbol{v}_i$ contains all DOF of the element T_i. Hence, the number of VDOF is $n_{vd} = 6$ for $d = 2$ and $n_{vd} = 12$ in the three-dimensional case. The algebraic edges on the fine level are naturally defined through $E = e_{ij} \in \mathcal{E}$ if $\overline{T}_i \cap \overline{T}_j \neq \emptyset$. Then, for any $E \in \mathcal{E}$ the edge matrices E_{ij} are naturally chosen to be A_E. That is, we have a direct decomposition of A without any approximation involved. Additionally, note that this splitting also takes care

[7]Their kernel is given by the rigid body motions.

of the boundary terms.

The detection of strong and weak connections via the strength of connectivity is determined exactly as in Definition 4.3.1 via the computational molecules $M(i,j)$ given by (4.23). Thereby, note that M_{ii} as well as M_{jj} is always SPD due to construction.

The set up of coarse edges aligns on the discussion of Section 4.4. We choose the set of coarse edges according to $\mathcal{A}_c = ((\mathcal{A}^s_{fc})^T \times \mathcal{A}^s_{fc}) \vee \mathcal{A}_{cc}$, from which we obtain the set of coarse edges $\mathcal{E}_c$. Now, for any $e_{ij} \in \mathcal{E}_c$ let $\mathcal{P}_{ij}$ denote the set of all algebraic edges which are contained in a strong path of length 2 connecting $\boldsymbol{v}_i$ and $\boldsymbol{v}_j$. If there exists a fine edge e_{ij} it is added to $\mathcal{P}_{ij}$ as well. Furthermore, let α_{kl} be the number how often the fine edge e_{kl} is used to set up a coarse edge, i.e., $\alpha_{kl} = |\{e_{ij} \in \mathcal{E}_c \,:\, e_{kl} \in \mathcal{P}_{ij}\}|$. Using the interpolation matrix P, the coarse edge matrix E^c_{ij} is set up via

$$E^c_{ij} = P^T \left(\sum_{e_{kl} \in \mathcal{P}_{ij}} \frac{1}{\alpha_{kl}} E_{kl} \right) P\,.$$

For the following discussion let us introduce the set of strong fine-to-coarse vertex connections $\mathcal{E}^s_{fc} := \{e_{kl} \in \mathcal{E} \,:\, \boldsymbol{v}_k \in \mathcal{V}_f \wedge \boldsymbol{v}_l \in \mathcal{V}_c \wedge s_{kl} \geq \theta\}$ and the set of coarse-to-coarse edges $\mathcal{E}_{cc} := \{e_{kl} \in \mathcal{E} \,:\, \boldsymbol{v}_k, \boldsymbol{v}_l \in \mathcal{V}_c\}$. Since our coarse-grid selection process meets the criterions (C1) and (C2) of Section 4.4 the following holds

$$\sum_{e_{ij} \in \mathcal{E}_c} E^c_{ij} = P^T \left(\sum_{e_{ij} \in \mathcal{E}_c} \sum_{e_{kl} \in \mathcal{P}_{ij}} \frac{1}{\alpha_{kl}} E_{kl} \right) P = P^T \left(\sum_{e_{ij} \in \mathcal{E}^s_{fc}} E_{ij} + \sum_{e_{ij} \in \mathcal{E}_{cc}} E_{ij} \right) P\,. \tag{4.40}$$

This means that the sum of all coarse edge matrices is equal to the Galerkin product of P with the sum of all strong fine-to-coarse and all coarse-to-coarse edge matrices. Therefore, the sum (4.40) is sparser than the coarse-grid operator $A_H = P^T A P$. This keeps the operator complexity lower on coarser levels, which is determined by the interpolation. The interpolation itself depends on the number of (strong and weak) connections of the algebraic mesh. It is set up exactly as discussed in Section 4.5. Finally, the smoothing is again chosen to be Gauss-Seidel smoothing.

4.9.2 Discussion

As it has been mentioned in the previous subsection, the number of VDOF is $n_{vd} = 6$ and $n_{vd} = 12$ for $d = 2$ and $d = 3$, respectively. This leads to edge matrices of the size 12×12

or 24×24. Those corresponding to interior edges without any Dirichlet boundary attached to one of the elements (see above) are SPSD. The other E_{ij} are SPD. That is, the general element matrix has rank 9 or 18, respectively. Hence, we have to store the whole matrix. In the following we neglect the symmetry of the problem, i.e, that all matrices can be stored more efficiently. Using the number of interior faces $n_{E,o} := |\mathcal{E}_h^o|$, we find that we require $12^2 n_{E,o}$ or $24^2 n_{E,o}$ of double precision memory entries for all edge matrices on one level. On the other hand, we obtain for the system matrix:

Lemma 4.9.1. *The system matrix A has in the case of full coupling between neighboring elements the following number of non-zero entries $\#nze(A)$:*

$$\#nze(A) = \begin{cases} 12n_E + 84n_{E,o} & d = 2\,, \\ 36n_E + 324n_{E,o} & d = 3\,, \end{cases} \tag{4.41}$$

where $n_{E,o} := |\mathcal{E}_h^o|$ is the number of interior faces.

Proof. Let us define the number of boundary faces $n_{E,\partial\Omega} := |\mathcal{E}_h^D \cup \mathcal{E}_h^N|$. First we consider the case $d = 2$. For every $T \in \mathcal{T}_h$ we have 6 DOF, hence, the total number of DOF is #DOF$= 6n_T$. Since every element T contains 3 edges we find $3n_T = 2n_E - n_{E,\partial\Omega}$, because all interior edges are counted twice and the boundary edges correspond only to one element. Every inner element has 3 adjacent $T \in \mathcal{T}_h$ leading to 24 non-zero entries for each DOF belonging to an interior element. For every boundary element, each DOF creates 24 non-zero entries minus 6 times the number of boundary edges the element belongs to. In total, we arrive at

$$\begin{aligned} \#\text{nze}(A) &= 24 \cdot 6n_T - 6 \cdot 6n_{E,\partial\Omega} = 96n_E - 48n_{E,\partial\Omega} - 36n_{E,\partial\Omega} \\ &= 96n_E - 84n_{E,\partial\Omega} = 12n_E + 84n_{E,o}\,. \end{aligned}$$

For $d = 3$ it holds that every element has 4 faces which leads to $4n_T = 2n_E - n_{E,\partial\Omega}$. In total there are $12n_T$ degrees of freedom. Now, every interior element has 12 DOF and 4 neighboring elements Hence, for each interior DOF A has 60 non-zero entries. Similar to the two-dimensional case, we find

$$\begin{aligned} \#\text{nze}(A) &= 60 \cdot 12n_T - 12 \cdot 12n_{E,\partial\Omega} = 360n_E - 180n_{E,\partial\Omega} - 144n_{E,\partial\Omega} \\ &= 360n_E - 324n_{E,\partial\Omega} = 36n_E + 324n_{E,o}\,. \end{aligned}$$

□

Hence, the memory consumption for setting up all E_{ij} is less than ($d = 2$) or less than two times ($d = 3$) the amount of memory we require for storing the system matrix A.

A real practical limitation is the size of the subproblems we have to solve for finding the strength of connectivities s_{ij}, see Definition 4.3.1. This procedure involves the solutions of generalized eigenvalue problems of size n_{vd}. Even though n_{vd} is relatively small, the total complexity for setting up all strengths of connectivity is still of order $\mathcal{O}(n_v)$ but with a high constant.

Finally, in the set up of the interpolation according to (4.29) we have to invert the matrix M_{ff}, which is of the size $n_{vd}(1 + |\mathcal{S}_i^f|)$, again leading to a very high constant in the complexity considerations.

In summary, we may conclude that the AMGm method will provide a uniform (with respect to number of DOFs and Poisson ratio) preconditioner for the DG discretization of linear elasticity equations. Clearly, achieving such robustness comes at the price of increasing the computational complexity of the algorithm. For values of the Poisson ratio that are far from the critical one, the AMGm method requires more computational work than the standard MG solvers for DG discretizations of elliptic problems, such as the ones developed in [AdDZ09, GK03].

Chapter 5

A subspace correction method for nearly singular linear elasticity problems

The focus of this chapter is on constructing a robust (uniform in the problem parameters) iterative solution method for the system of linear algebraic equations arising from a nonconforming finite element discretization based on reduced integration. We introduce a specific space decomposition into two overlapping subspaces that serves as a basis for devising a uniformly convergent subspace correction algorithm. We consider the equations of linear elasticity in primal variables. For nearly incompressible materials, i.e., when the Poisson ratio ν approaches $1/2$, this problem becomes ill-posed and the resulting discrete problem is nearly singular.

Subspace correction methods for nearly singular systems have been studied in [LWXZ07]. Therein the authors show that a space decomposition has to fulfill a local kernel preservation property in order to set up a robust MSSC on top of this splitting. This leads to robust multigrid methods for planar linear elasticity problems (see [LWC09]). In [Sch99a, Sch99b] a multigrid method has been presented for a finite element discretization with $P_2 - P_0$ elements. This approach relies on a local basis for the weakly divergence-free functions.

In this setting, presently known (multilevel) iterative solution methods are optimal or nearly optimal for the pure displacement problem only, i.e., when Dirichlet boundary conditions are imposed on the entire boundary, see, e.g., [BS07, GKM10]. For pure traction or mixed boundary conditions the problem gets more involved. It is known, that standard (conforming and nonconforming) finite element methods then require certain stabilization techniques, see, e.g., [HL03, Fal91]. We employ the discretization scheme introduced in Subsection 2.4.2, which achieves the stabilization via reduced integration. Note that optimal error estimates have been shown for this method (see [Fal91]).

The remainder of this chapter is organized as follows: The problem formulation of the linear elasticity problem with pure traction boundary conditions and its finite element discretization are recalled in Section 5.1. Additionally, the necessary ingredients for the discussion in this chapter are introduced. In Section 5.2 we present a specific space decomposition, with newly derived, locally supported basis functions, which is the basis for an MSSC solver being discussed in Section 5.3. Afterwards, we introduce the preconditioner that is naturally defined by the MSSC. In Section 5.5 it is explained how to solve the subproblems efficiently. Finally, we present numerical tests illustrating the optimal performance of the preconditioner in Section 5.6. Additionally, results for the direct application of the MSSC are shown.

5.1 Preliminaries

For the sake of simplicity we consider only two-dimensional problems ($d = 2$) in this chapter. Let us focus on problem (2.33), with $a(.,.)$ and $L(.)$ given by (2.52) and (2.53). The space V_h is given by $\hat{\boldsymbol{V}}_h$ according to Subsection 2.4.2. This problem is stable with respect to the first Lamé parameter and therefore, it may be exploited to investigate almost incompressible materials. In this chapter, we use the notation introduced in Subsection 2.4.2.

Let $\mathcal{E}_H$ be the set of edges of $\mathcal{T}_H$ and $\mathcal{V}_H$ be the set of (coarse) vertices of the mesh $\mathcal{T}_H$. Then for any vertex $v_i \in \mathcal{V}_H$ we denote the set of edges sharing v_i by $\mathcal{N}_i^{\mathcal{E}}$, and by $\mathcal{N}_i^{\mathcal{T}_H}$ we designate the set of elements $T \in \mathcal{T}_H$ for which v_i is a vertex. The set $\mathcal{E}_T$ contains all edges of an element $T \in \mathcal{T}_H$. For any edge $E = (v_{E,1}, v_{E,2}) \in \mathcal{E}_H$ by φ_E we designate the scalar nodal basis function of V_h^1 corresponding to the midpoint of the edge E, and by $\varphi_{E,1}$ and $\varphi_{E,2}$ the nodal basis functions corresponding to the vertices $v_{E,1}$ and $v_{E,2}$ of E. The corresponding vector-valued DOFs of any function $\boldsymbol{v}_h \in \boldsymbol{V}_h := [V_h^1]^d$ are labeled by $\boldsymbol{v}_E$, $\boldsymbol{v}_{E,1}$ and $\boldsymbol{v}_{E,2}$, respectively. We further use φ_i and $\boldsymbol{v}_i$ to denote the basis functions and DOFs associated with the vertices from $\mathcal{V}_H$.

For any edge $E \in \mathcal{E}_H$ we assume that $v_{E,1} < v_{E,2}$ and that the globally defined tangential vector $\boldsymbol{\tau}_E$ points from $v_{E,1}$ to $v_{E,2}$. The global edge normal vector $\boldsymbol{n}_E$ is orthogonal to $\boldsymbol{\tau}_E$ and is obtained from τ_E by a clockwise rotation. Especially, we need the space $\boldsymbol{V}_{H,0}^{RT}$ of lowest order Raviart Thomas functions (2.32). The basis functions $\boldsymbol{\varphi}_E^{RT}$ corresponding to an edge E of an element $T \in \mathcal{T}_H$ are such that $N_{E'}^{RT}(\boldsymbol{\varphi}_E^{RT}) = \delta_{EE'}$, according to the definition of a nodal basis in Subsection 2.1.3. We also use the projection $\Pi^{RT} : H(\mathrm{div};\Omega) \mapsto \boldsymbol{V}_{H,0}^{RT}$ defined by

$$\Pi^{RT}(\boldsymbol{v}) = \sum_{E\in\mathcal{E}_H} N_E^{RT}(\boldsymbol{v})\boldsymbol{\varphi}_E^{RT}\,, \tag{5.1}$$

for which the commuting property $P_0 \,\mathrm{div}\, \boldsymbol{v} = \mathrm{div}\, \Pi^{RT}(\boldsymbol{v})$ holds for any $\boldsymbol{v} \in H(\mathrm{div};\Omega)$ (cf. [BF91, p. 131]).

Lemma 5.1.1. *For any* $\boldsymbol{v} \in H(\mathrm{div};\Omega)$ *we have*

$$P_0 \,\mathrm{div}\, \boldsymbol{v} = \mathrm{div}\, \Pi^{RT}(\boldsymbol{v})\,. \tag{5.2}$$

In [BF91] Lemma 5.1.1 is proven. For a better understanding we add a proof of the previous lemma for functions $\boldsymbol{v}_h \in \boldsymbol{V}_h$.

Proof. Let us consider a single triangle $T \in \mathcal{T}_H$ and let us denote by $\boldsymbol{n}_{E,T}$ the local unit normal vector pointing outwards on T. Then for any $\boldsymbol{v}_h \in \boldsymbol{V}_h$ we have

$$\begin{aligned}
\mathrm{div}\,\Pi^{RT}(\boldsymbol{v}_h)\big|_T &= \mathrm{div}\left(\sum_{E\in\mathcal{E}_T} N_E^{RT}(\boldsymbol{v}_h)\boldsymbol{\varphi}_E^{RT}\right) = \sum_{E\in\mathcal{E}_T} N_E^{RT}(\boldsymbol{v}_h)\,\mathrm{div}\,\boldsymbol{\varphi}_E^{RT} \\
&= \sum_{E\in\mathcal{E}_T} \frac{(\boldsymbol{n}_E\cdot\boldsymbol{n}_{E,T})}{|E|}\int_E \boldsymbol{v}_h\cdot\boldsymbol{n}_E\,\mathrm{d}s\,\frac{|E|}{|T|} = \frac{1}{|T|}\int_{\partial T}\boldsymbol{v}_h\cdot\boldsymbol{n}_{E,T}\,\mathrm{d}s \\
&= \frac{1}{|T|}\int_T \mathrm{div}\,\boldsymbol{v}_h\,\mathrm{d}\boldsymbol{x} = P_0(\mathrm{div}\,\boldsymbol{v}_h)|_T\,,
\end{aligned}$$

by using the relation $\mathrm{div}\,\boldsymbol{\varphi}_E^{RT} = (\boldsymbol{n}_E\cdot\boldsymbol{n}_{E,T})\frac{|E|}{|T|}$, which follows from

$$\begin{aligned}
|T|\,\mathrm{div}\,\boldsymbol{\varphi}_E^{RT} &= \int_T \mathrm{div}\,\boldsymbol{\varphi}_E^{RT}\,\mathrm{d}\boldsymbol{x} = \sum_{E'\in\mathcal{E}_T}\int_{E'}\boldsymbol{\varphi}_E^{RT}\cdot\boldsymbol{n}_{E',T}\,\mathrm{d}s \\
&= (\boldsymbol{n}_E\cdot\boldsymbol{n}_{E,T})\int_E \boldsymbol{\varphi}_E^{RT}\cdot\boldsymbol{n}_E\,\mathrm{d}s = (\boldsymbol{n}_E\cdot\boldsymbol{n}_{E,T})|E|\,,
\end{aligned}$$

since $\mathrm{div}\,\boldsymbol{\varphi}_E^{RT}$ is constant on an element $T \in \mathcal{T}_H$. Summation over all elements yields the desired result. □

Additionally, let us designate by $n_{E,H}$, $n_{T,H}$ and $n_{V,H}$ the cardinalities of the sets $\mathcal{E}_H$, $\mathcal{T}_H$ and $\mathcal{V}_H$. That is, those values refer to the number of coarse edges, coarse elements (triangles) and to the number of coarse vertices.

5.2 Space decomposition

In the following we present innovative results that have not been published before by another author. Let us consider the following unique decomposition of any function $\boldsymbol{v}_h \in \boldsymbol{V}_h$:

$$\begin{aligned}
\boldsymbol{v}_h &= \sum_{v_i \in \mathcal{V}_H} \varphi_i \boldsymbol{v}_i + \sum_{E \in \mathcal{E}_H} \varphi_E \boldsymbol{v}_E \\
&= \underbrace{\sum_{v_i \in \mathcal{V}_H} \Big[\varphi_i \boldsymbol{v}_i - \frac{1}{2} \sum_{E \in \mathcal{N}_i^{\mathcal{E}}} (\boldsymbol{v}_i \cdot \boldsymbol{n}_E) \varphi_E \boldsymbol{n}_E\Big]}_{=:\boldsymbol{v}_{\mathcal{V}}} + \underbrace{\sum_{E \in \mathcal{E}_H} (\boldsymbol{v}_E \cdot \boldsymbol{\tau}_E) \varphi_E \boldsymbol{\tau}_E}_{=:\boldsymbol{v}_\tau} \\
&\quad \underbrace{+ \sum_{E \in \mathcal{E}_H} \left(\left[\boldsymbol{v}_E + \frac{1}{2}(\boldsymbol{v}_{E,1} + \boldsymbol{v}_{E,2}) \right] \cdot \boldsymbol{n}_E \right) \varphi_E \boldsymbol{n}_E}_{=:\boldsymbol{v}_1} .
\end{aligned}$$

Next we define the splitting $\boldsymbol{V}_h = \boldsymbol{V}_{\mathcal{V}} \oplus \boldsymbol{V}_\tau \oplus \boldsymbol{V}_n$, where

$$\boldsymbol{V}_{\mathcal{V}} := \{\boldsymbol{v}_h \in \boldsymbol{V}_h : \boldsymbol{v}_h = \sum_{v_i \in \mathcal{V}_H} \Big[\varphi_i \boldsymbol{v}_i - \frac{1}{2} \sum_{E \in \mathcal{N}_i^{\mathcal{E}}} (\boldsymbol{v}_i \cdot \boldsymbol{n}_E) \varphi_E \boldsymbol{n}_E\Big]\} , \tag{5.3}$$

$$\boldsymbol{V}_\tau := \{\boldsymbol{v}_h \in \boldsymbol{V}_h : \boldsymbol{v}_h = \sum_{E \in \mathcal{E}_H} \alpha_E \varphi_E \boldsymbol{\tau}_E\} , \tag{5.4}$$

$$\boldsymbol{V}_n := \{\boldsymbol{v}_h \in \boldsymbol{V}_h : \boldsymbol{v}_h = \sum_{E \in \mathcal{E}_H} \alpha_E \varphi_E \boldsymbol{n}_E\} . \tag{5.5}$$

Note that $\Pi^{RT}(\boldsymbol{V}_{\mathcal{V}}) = \Pi^{RT}(\boldsymbol{V}_\tau) = \{0\}$. In addition, let us introduce the spaces

$$\boldsymbol{V}_{\text{curl}} := \{\boldsymbol{v}_h \in \boldsymbol{V}_h : \boldsymbol{v}_h = \sum_{v_i \in \mathcal{V}_H} \beta_i \sum_{E \in \mathcal{N}_i^{\mathcal{E}}} \frac{\delta_{E,i}}{|E|} \varphi_E \boldsymbol{n}_E\} , \tag{5.6}$$

$$\boldsymbol{V}_{\nabla_h} := \{\boldsymbol{v}_h \in \boldsymbol{V}_h : \boldsymbol{v}_h = \sum_{T \in \mathcal{T}_H} \gamma_T \sum_{E \in \mathcal{E}_T} (\boldsymbol{n}_E \cdot \boldsymbol{n}_{E,T}) \varphi_E \boldsymbol{n}_E\} . \tag{5.7}$$

Here $\delta_{E,i}$ is defined by

$$\delta_{E,i} = \begin{cases} -1 & \text{if} \quad i = v_{E,1} \\ 1 & \text{if} \quad i = v_{E,2} \end{cases} . \tag{5.8}$$

Note that $\boldsymbol{V}_{\text{curl}} \subset \boldsymbol{V}_n$, and $\boldsymbol{V}_{\nabla_h} \subset \boldsymbol{V}_n$, and the following properties hold:

Lemma 5.2.1. *The space $\boldsymbol{V}_{\text{curl}}$ is "weakly divergence-free", while $\boldsymbol{V}_{\nabla_h}$ is its complement, i.e.,*

$$P_0 \operatorname{div}(\boldsymbol{v}_{\text{curl}}) = \operatorname{div} \Pi^{RT}(\boldsymbol{v}_{\text{curl}}) = 0 \qquad \forall \boldsymbol{v}_{\text{curl}} \in \boldsymbol{V}_{\text{curl}} , \tag{5.9}$$

$$P_0 \operatorname{div}(\boldsymbol{v}_{\nabla_h}) = \operatorname{div} \Pi^{RT}(\boldsymbol{v}_{\nabla_h}) \neq 0 \qquad \forall \boldsymbol{v}_{\nabla_h} \in \boldsymbol{V}_{\nabla_h} . \tag{5.10}$$

Furthermore, it holds that $\dim(\boldsymbol{V}_{\text{curl}}) = n_{v,H} - 1$ *and* $\dim(\boldsymbol{V}_{\nabla_h}) = n_{T,H}$*, and thus, we find* $\boldsymbol{V}_n = \boldsymbol{V}_{\text{curl}} \oplus \boldsymbol{V}_{\nabla_h}$.

Proof. We observe that $\Pi^{RT}(\varphi_E \boldsymbol{n}_E) = \frac{1}{2}\boldsymbol{\varphi}_E^{RT}$ for any $E \in \mathcal{E}_H$.

In the proof of Lemma 5.1.1, we have seen, that $\operatorname{div} \boldsymbol{\varphi}_E^{RT} = (\boldsymbol{n}_E \cdot \boldsymbol{n}_{E,T})\frac{|E|}{|T|}$. Now, let for any $\boldsymbol{v}_{RT} \in \boldsymbol{V}_{H,0}^{RT}$, i.e., $\boldsymbol{v}_{RT} = \sum_{E\in\mathcal{E}_H} \alpha_E \boldsymbol{\varphi}_E^{RT}$, be $\boldsymbol{\alpha} = (\alpha_E)_{E\in\mathcal{E}_H} \in \mathbb{R}^{n_{E,H}}$ the corresponding vector of DOFs to $\boldsymbol{v}_{RT}$. Since $\operatorname{div} \boldsymbol{v}_{RT}$ is piecewise constant, it is given by $\operatorname{div} \boldsymbol{v}_{RT} = \sum_{T\in\mathcal{T}_H} \beta_T \chi_T$ with $\boldsymbol{\beta} = (\beta_T)_{T\in\mathcal{T}_H}$ specifying its vector of DOFs. In the latter statement χ_T denotes the characteristic function of an element $T \in \mathcal{T}_H$. For each $T \in \mathcal{T}_H$ we obtain

$$\beta_T = \sum_{E\in\mathcal{E}_T} \alpha_E \operatorname{div} \boldsymbol{\varphi}_E^{RT}\big|_T = \sum_{E\in\mathcal{E}_T} \alpha_E (\boldsymbol{n}_E \cdot \boldsymbol{n}_{E,T})\frac{|E|}{|T|}\,.$$

In the following, let $D_{\mathcal{E}_H} : \mathbb{R}^{n_{E,H}} \to \mathbb{R}^{n_{E,H}}$ and $D_{\mathcal{T}_H} : \mathbb{R}^{n_{T,H}} \to \mathbb{R}^{n_{T,H}}$ be the diagonal matrices of face (edge) and element measures, i.e., $D_{\mathcal{E}_H} = \operatorname{diag}(|E|)$ and $D_{\mathcal{T}_H} = \operatorname{diag}(|T|)$. Additionally, let $B : \mathbb{R}^{n_{E,H}} \to \mathbb{R}^{n_{T,H}}$ be defined by $B = (B_{TE})_{E\in\mathcal{E}_H, T\in\mathcal{T}_H}$ where

$$B_{TE} := \begin{cases} 0 & \text{if } \; E \notin \mathcal{E}_T \\ \boldsymbol{n}_{E,T} \cdot \boldsymbol{n}_E & \text{if } \; E \in \mathcal{E}_T \end{cases}.$$

With those matrices, we find that

$$\boldsymbol{\beta} = D_{\mathcal{T}_H}^{-1} B D_{\mathcal{E}_H} \boldsymbol{\alpha}\,.$$

That is, $\operatorname{div} \boldsymbol{v}_{RT} = 0$ if and only if $\boldsymbol{\beta} = \boldsymbol{0}$. Hence, we have to show that for every $\boldsymbol{\alpha}$ corresponding to a basis function of $\boldsymbol{V}_{\text{curl}}$ the relation $\boldsymbol{\beta} = D_{\mathcal{T}_H}^{-1} B D_{\mathcal{E}_H} \boldsymbol{\alpha} = \boldsymbol{0}$ is valid. Let us consider $\boldsymbol{\alpha}_i$ being the vector representation of a basis function in $\boldsymbol{V}_{\text{curl}}$ of an arbitrary vertex $v_i \in \mathcal{V}_H$. For any $T \notin \mathcal{N}_i^{\mathcal{T}_H}$ we immediately see that $\beta_T = 0$. For all other T, $\boldsymbol{\alpha}$ contains two non-vanishing values. They belong to the edges $E_1, E_2 \in \mathcal{E}_T$ being adjacent to the actually considered vertex v_i. We find that $\beta_T = |T|(\boldsymbol{n}_{E_1} \cdot \boldsymbol{n}_{E_1,T}\delta_{E_1,i} + \boldsymbol{n}_{E_2} \cdot \boldsymbol{n}_{E_2,T}\delta_{E_2,i})$. Due to the special choice of $\delta_{E,i}$ and because $\boldsymbol{n}_{E_1,T}$ as well as $\boldsymbol{n}_{E_2,T}$ point outwards of T, β_T vanishes. Hence $\boldsymbol{\beta} = \boldsymbol{0}$ follows, which implies statement (5.9).

In order to show (5.10) first note that for every $T \in \mathcal{T}_H$ we find

$$\begin{aligned} \operatorname{div} \sum_{E\in\mathcal{E}_T} (\boldsymbol{n}_E \cdot \boldsymbol{n}_{E,T})\boldsymbol{\varphi}_E^{RT} &= \sum_{E\in\mathcal{E}_T} (\boldsymbol{n}_E \cdot \boldsymbol{n}_{E,T}) \operatorname{div} \boldsymbol{\varphi}_E^{RT} \\ &= \sum_{E\in\mathcal{E}_T} (\boldsymbol{n}_E \cdot \boldsymbol{n}_{E,T})^2 \frac{|E|}{|T|} = \sum_{E\in\mathcal{E}_T} \frac{|E|}{|T|} \neq 0\,. \end{aligned}$$

Next we show that the $\boldsymbol{\alpha}$ corresponding to the basis functions of $\boldsymbol{V}_{\nabla_h}$ are linearly independent, which implies that $\operatorname{div} \boldsymbol{v}_{\nabla_h} \neq 0$ for all $\boldsymbol{v}_{\nabla_h} \in \boldsymbol{V}_{\nabla_h}$. Therefore, note that all $\boldsymbol{\alpha}_T$ corresponding to an element $T \in \mathcal{T}_H$ on the boundary of the domain are linearly independent because they are the only ones to contain an entry for the boundary edges. If we continue with this argument, excluding the vectors already treated, we conclude that all $\boldsymbol{\alpha}_T$ are indeed linearly independent.

This implies that B has full rank, i.e., $\dim \boldsymbol{V}_{\nabla_h} = n_{T,H}$ and on the other hand that $\dim \boldsymbol{V}_{\text{curl}} = n_{V,H} - 1$, since $n_{E,H} = n_{T,H} + n_{V,H} - 1$ due to Euler's formula. From (5.6) we find $n_{V,H}$ possible basis functions of $\boldsymbol{V}_{\text{curl}}$. Actually, one may prove with similar arguments as above, that $n_{V,H} - 1$ of them, each corresponding to a vertex $v_i \in \mathcal{V}_H$, are linearly independent. Summarizing we have shown that $\boldsymbol{V}_n = \boldsymbol{V}_{\text{curl}} \oplus \boldsymbol{V}_{\nabla_h}$. □

Lemma 5.2.2.

$$\boldsymbol{V}_h = \boldsymbol{V}_{\mathcal{V}} \oplus \boldsymbol{V}_{\tau} \oplus \boldsymbol{V}_{\text{curl}} \oplus \boldsymbol{V}_{\nabla_h}. \tag{5.11}$$

Proof. Clearly, from (5.3) we see that $\dim \boldsymbol{V}_{\mathcal{V}} = 2n_{V,H}$ corresponding to the coarse vertices $\mathcal{V}_H$. Moreover, for any vertex $E \in \mathcal{E}_H$ the vectors $\boldsymbol{\tau}_E$ and $\boldsymbol{n}_E$ are linearly independent. Hence, $\boldsymbol{V}_n \cap \boldsymbol{V}_{\tau} = \emptyset$ which in the end implies (5.11). □

We want to decompose the space $\boldsymbol{V}_h$ into two overlapping subspaces such that the problem on the subspaces is somehow "easier" solvable than on the whole space. Additionally, in Section 3.6 we have seen that a proper choice of the overlap is crucial for the convergence performance of the MSSC method. Numerical experiments show that the CBS constant between $\boldsymbol{V}_{\mathcal{V}}$ and $\boldsymbol{V}_{\nabla_h}$ is bounded away from 1. Hence, we decompose $\boldsymbol{V}_h$ into two overlapping subspaces $\boldsymbol{V}_I$ and $\boldsymbol{V}_{II}$:

$$\boldsymbol{V}_I = \boldsymbol{V}_{\mathcal{V}} \oplus \boldsymbol{V}_{\tau} \oplus \boldsymbol{V}_{\text{curl}} \tag{5.12}$$

$$\boldsymbol{V}_{II} = \boldsymbol{V}_{\tau} \oplus \boldsymbol{V}_{\text{curl}} \oplus \boldsymbol{V}_{\nabla_h} \tag{5.13}$$

The overlap of $\boldsymbol{V}_I$ and $\boldsymbol{V}_{II}$ is given by $\boldsymbol{V}_{\tau} \oplus \boldsymbol{V}_{\text{curl}}$, and any element $\boldsymbol{v}_{II} \in \boldsymbol{V}_{II}$ can be uniquely decomposed into $\boldsymbol{v}_{II} = \boldsymbol{v}_{\tau} + \boldsymbol{v}_{\text{curl}} + \boldsymbol{v}_{\nabla_h}$, with $\boldsymbol{v}_{\tau} \in \boldsymbol{V}_{\tau}$, $\boldsymbol{v}_{\text{curl}} \in \boldsymbol{V}_{\text{curl}}$ and $\boldsymbol{v}_{\nabla_h} \in \boldsymbol{V}_{\nabla_h}$. However, finding the components $\boldsymbol{v}_{\text{curl}} \in \boldsymbol{V}_{\text{curl}}$ and $\boldsymbol{v}_{\nabla_h} \in \boldsymbol{V}_{\nabla_h}$ for a given function $\boldsymbol{v}_n \in \boldsymbol{V}_n$ requires a solution of a system with an M-matrix corresponding to the lowest order mixed method for the Laplace equation with lumped mass [BF91].

Note that since $P_0 \operatorname{div}(\boldsymbol{V}_I) = \operatorname{div} \Pi^{RT}(\boldsymbol{V}_I) = \{0\}$ the bilinear form $a(.,.)$ satisfies

$$a(\boldsymbol{u}_I, \boldsymbol{v}_I) = 2\mu(\varepsilon(\boldsymbol{u}_I), \varepsilon(\boldsymbol{v}_I))_0 \qquad \forall \boldsymbol{u}_I, \boldsymbol{v}_I \in \boldsymbol{V}_I. \tag{5.14}$$

On the other hand, in the limit case $\nu = 0$ we have $a(\boldsymbol{u}_h, \boldsymbol{v}_h) = 2\mu(\varepsilon(\boldsymbol{u}_h), \varepsilon(\boldsymbol{v}_h))_0$ for all $\boldsymbol{u}_h, \boldsymbol{v}_h \in \boldsymbol{V}_h$.

5.3 MSSC as a solver

The bilinear form $2\mu(\varepsilon(\boldsymbol{u}_h), \varepsilon(\boldsymbol{v}_h))_0$ is spectrally equivalent to the vector Laplace equation because of Korn's inequality. It is well-known how to solve or precondition this term efficiently. Therefore, we add another correction step and choose $\boldsymbol{V}_{III} = \boldsymbol{V}_h$ and $a_{III}(.,.) := 2\mu(\varepsilon(.), \varepsilon(.))_0$, with its operator representation $2\mu A_\varepsilon$. Let A denote the operator corresponding to the bilinear form $a(.,.)$ given by (2.52). In order to get a non-expansive error propagation operator we add the update with a suitable factor $c_{III}(\nu)$. Now, this MSSC is defined through

$$E_{\text{MSSC}} = (I - \frac{c_{III}(\nu)}{2\mu} A_\varepsilon^{-1} A)(I - P_{II})(I - P_I) \tag{5.15}$$

where P_i is defined by (3.15) for $i = I, II$. The factor $c_{III}(\nu)$ can be defined by

$$c_{III}(\nu) := \frac{1}{c_{III}} \min\{1, \frac{1-2\nu}{\nu}\}, \tag{5.16}$$

where c_{III} is chosen properly depending on the maximal eigenvalue of $A_\varepsilon^{-1}A$. In order to assure that the operator $I - \frac{c_{III}(\nu)}{2\mu} A_\varepsilon^{-1}A$ is non-expansive we require

$$0 \leq \frac{c_{III}(\nu)}{2\mu} A_\varepsilon^{-1} A \leq 2I .$$

With the operator form A_{div} corresponding to $(P_0 \operatorname{div} ., P_0 \operatorname{div} .)_0$ we arrive at

$$\frac{c_{III}(\nu)}{\mu} \left(\mu I + \frac{\mu\nu}{1-2\nu} A_\varepsilon^{-1} A_{\text{div}} \right) \leq 2I . \tag{5.17}$$

Overall, we find that the operator is non-expansive for all $\nu \in [0, 1/2]$, if c_{III} fulfills

$$c_{III} \geq \frac{1}{2} \left(1 + \lambda_{\max}(A_\varepsilon^{-1} A_{\text{div}})\right) . \tag{5.18}$$

Finally, we estimate $\lambda_{\max}(A_\varepsilon^{-1} A_{\text{div}})$.

Lemma 5.3.1. *Let $d = 2$, A_ε and A_{div} be the operators corresponding to the bilinear forms $(\varepsilon(.), \varepsilon(.))_0$ and $(P_0 \operatorname{div} ., P_0 \operatorname{div} .)_0$, with respect to the space $\hat{\boldsymbol{V}}_h$. Then, we find*

$$\lambda_{\max}(A_\varepsilon^{-1} A_{\text{div}}) \leq \frac{2}{c_K} \tag{5.19}$$

with c_K being the constant in the Korn inequality (Theorem 2.1.6).

Proof. First, note that $\operatorname{div} \boldsymbol{v}_h \in S_h$ and hence, for arbitrary $s_h \in S_h$ we have

$$\begin{aligned}
\|P_0 s_h\|_0^2 &= \int_\Omega \left(\sum_{T_H \in \mathcal{T}_H} \chi_{T_H}(\boldsymbol{y}) \frac{1}{|T_H|} \int_{T_H} s_h(\boldsymbol{x})\,\mathrm{d}\boldsymbol{x} \right)^2 \mathrm{d}\boldsymbol{y} \\
&= \sum_{T_H' \in \mathcal{T}_H} \int_{T_H'} \left(\sum_{T_H \in \mathcal{T}_H} \chi_{T_H} \frac{1}{4} \sum_{T_h \subset T_H} s_{h,T_h} \right)^2 \mathrm{d}\boldsymbol{x} \\
&= \frac{1}{16} \sum_{T_H \in \mathcal{T}_H} \int_{T_H} \left(\sum_{T_h \subset T_H} s_{h,T_h} \right)^2 \mathrm{d}\boldsymbol{x} \\
&\overset{\text{C.S.}}{\leq} \frac{1}{16} \sum_{T_H \in \mathcal{T}_H} |T_H| 4 \sum_{T_h \subset T_H} s_{h,T_h}^2 = \sum_{T_H \in \mathcal{T}_H} \sum_{T_h \subset T_H} |T_h| s_{h,T_h}^2 = \|s_h\|_0^2 , \qquad (5.20)
\end{aligned}$$

where we have used Cauchy-Schwarz inequality (C.S.). Thus, we find with Cauchy-Schwarz inequality and Theorem 2.1.6 for arbitrary $\boldsymbol{v}_h \in \hat{\boldsymbol{V}}_h$

$$\begin{aligned}
\|P_0 \operatorname{div} \boldsymbol{v}_h\|_0^2 &\overset{(5.20)}{\leq} \|\operatorname{div} \boldsymbol{v}_h\|_0^2 = \int_\Omega \left(\sum_{i=1}^2 \frac{\partial u_i}{\partial x_i} \right)^2 \mathrm{d}\boldsymbol{x} \overset{\text{C.S.}}{\leq} 2 \sum_{i=1}^2 \int_\Omega \left(\frac{\partial u_i}{\partial x_i} \right)^2 \mathrm{d}\boldsymbol{x} \\
&\leq 2\|\boldsymbol{\nabla} \boldsymbol{v}_h\|_0^2 \leq 2\|\boldsymbol{v}_h\|_1^2 \overset{\text{Theorem 2.1.6}}{\leq} \frac{2}{c_k} \|\boldsymbol{\varepsilon}(\boldsymbol{v}_h)\|_0^2 .
\end{aligned}$$

Finally, we arrive at

$$\lambda_{\max}(A_\varepsilon^{-1} A_{\text{div}}) = \sup_{\boldsymbol{v}_h \in \hat{\boldsymbol{V}}_h} \frac{(A_{\text{div}} \boldsymbol{v}_h , \boldsymbol{v}_h)_0}{(A_\varepsilon \boldsymbol{v}_h , \boldsymbol{v}_h)_0} = \sup_{\boldsymbol{v}_h \in \hat{\boldsymbol{V}}_h} \frac{\|P_0 \operatorname{div} \boldsymbol{v}_h\|_0}{\|\boldsymbol{\varepsilon}(\boldsymbol{v}_h)\|_0} \leq \frac{2}{c_k} .$$

□

The latter lemma implies that if c_{III} is chosen such that

$$\frac{1}{2}\left(1 + \lambda_{\max}(A_\varepsilon^{-1} A_{\text{div}})\right) \leq \frac{1}{2}\left(1 + \frac{2}{c_k}\right) \leq c_{III} ,$$

it is guaranteed that E_{MSSC} is non-expansive. In order to assure that such a property holds one needs to have the precise value of the Korn's constant for the domain Ω. In general, this constant is not known and to make the method practical we need to modify it so that this unknown constant is not used in the algorithm. In order to do this we will relax the requirement that the iteration is a convergent one and in the next section we show how to use the multiplicative method as a preconditioner.

5.4 MSSC as a preconditioner

The error propagation operator E_{MSSC}, given by (5.15), automatically defines a convergent procedure if condition (5.17) is fulfilled. If this condition is satisfied with even stronger bounds, i.e., some uniform bounds away from 0 and 2, we would have a uniformly convergent method. Thereby, as we have explained above, a serious problem is the right choice of c_{III}. Since its estimate depends on Korn's constant it is very hard to meet the criterion in general.

However, the upper bound $2I$ can be relaxed. If we would have instead of (5.17) a condition like

$$0 < \alpha_0 I \leq \frac{c_{III}(\nu)}{\mu}\left(\mu I + \frac{\mu\nu}{1-2\nu} A_\varepsilon^{-1} A_{\text{div}}\right) \leq \alpha_1 I\,, \tag{5.21}$$

for some $0 < \alpha_0 < \alpha_1$, E_{MSSC} would naturally define a uniform unsymmetric preconditioner B_{MSSC}, through

$$E_{\text{MSSC}} = I - B_{\text{MSSC}} A\,.$$

Therefore, let us start with the MSSC for $J = 2$, i.e., the spaces $\boldsymbol{V}_I$ and $\boldsymbol{V}_{II}$ defined by (5.12) and (5.13), respectively. If we symmetrize this procedure, we obtain the following error propagation $\bar{E}_{\text{MSSC}}$, compare with (3.7) in case of $J = 2$ and exact subsolves, i.e.,

$$\bar{E}_{\text{MSSC}} = (I - P_I)(I - P_{II})(I - P_I)\,.$$

The error propagation operator can be rewritten as $\bar{E}_{\text{MSSC}} = I - \bar{B}_{\text{MSSC}} A$, with symmetric $\bar{B}_{\text{MSSC}}$. Further, $\bar{B}_{\text{MSSC}}$ is positive definite, since $\bar{E}_{\text{MSSC}}$ is non-expansive. Note that even though $\bar{B}_{\text{MSSC}} = (I - \bar{E}_{\text{MSSC}})A^{-1}$ formally involves the inverse of A, we do not need A^{-1} in order to apply $\bar{B}_{\text{MSSC}}$.

If ν is bounded away from the incompressible limit $1/2$, we know that A_ε is spectrally equivalent to A. Further, there are efficient preconditioners for A_ε. Therefore, w define the additive preconditioner B by

$$B := \frac{1-2\nu}{1-\nu} A_\varepsilon^{-1} + \frac{\nu}{1-\nu} \bar{B}_{\text{MSSC}}\,. \tag{5.22}$$

Note that B is a convex combination of A_ε^{-1} and $\bar{B}_{\text{MSSC}}$.

5.5 Solution of the subproblems

In order to solve the problem (2.33), either by means of Algorithm 3.1.1 in the setting of Section 5.3 or by using the preconditioner (5.22) of Section 5.4 in a PCG iteration, we have to solve the three subproblems. So far, we have assumed to solve those problems exactly. However, from the results in [Xu92, XZ02] we know that under certain conditions an inexact solution of problem (3.4) on each subspace results in a uniform preconditioner. Equivalently, if the conditions on T_i of Theorem 3.1.1 are fulfilled uniformly, the method according to the error propagation (5.15) is uniformly convergent.

The subproblems on the spaces $\boldsymbol{V}_I$ and $\boldsymbol{V}_{III} = \boldsymbol{V}_h$ involve the bilinear form

$$\tilde{a}(\boldsymbol{u}_i, \boldsymbol{v}_i) = 2\mu(\varepsilon(\boldsymbol{u}_i), \varepsilon(\boldsymbol{v}_i))_0 \qquad \forall \boldsymbol{u}_i, \boldsymbol{v}_i \in \boldsymbol{W} = \boldsymbol{V}_I, \boldsymbol{V}_h . \tag{5.23}$$

Any efficient preconditioning technique for the vector Laplace equation may be employed in these steps, e.g., classical AMG (see [RS87]) or AMGm (see Chapter 4 or [KK10]).

The problem on $\boldsymbol{V}_{II} = \boldsymbol{V}_E := \{\boldsymbol{v}_h \in \boldsymbol{V}_h \; : \; \boldsymbol{v}_h(\boldsymbol{x}_i) = \boldsymbol{0} \;\; v_i \in \mathcal{V}_H\}$ is more involved. In order to solve it one can exploit two equivalence relations. Intuitively, one might think it is obvious that Korn's inequality holds on $\boldsymbol{V}_E$ because of the high oscillatory behavior of functions in $\boldsymbol{V}_E$. Nevertheless, we provide a proof in the next theorem. The proof follows the proof of Theorem 11.2.16 in [BS07].

Theorem 5.5.1 (Korn's inequality on $\boldsymbol{V}_E$). *There exists a positive constant $c_K > 0$ such that*

$$\|\varepsilon(\boldsymbol{v}_E)\|_{L^2(\Omega)} \geq c_K \|\boldsymbol{v}_E\|_{H^1(\Omega)} \qquad \forall \boldsymbol{v}_E \in \boldsymbol{V}_E . \tag{5.24}$$

Proof. Due to Lemma 2.1.2 every $\boldsymbol{v}_E \in \boldsymbol{V}_E$ can be uniquely decomposed into $\boldsymbol{v}_E = \hat{\boldsymbol{v}} + \boldsymbol{v}_{\mathrm{RBM}}$ with $\hat{\boldsymbol{v}} \in \hat{\boldsymbol{H}}^1(\Omega)$ and $\boldsymbol{v}_{\mathrm{RBM}} \in \boldsymbol{V}^{\mathrm{RBM}}$. Now, Lemma 2.1.8 implies the existence of a $C > 0$ such that

$$C\left(\|\hat{\boldsymbol{v}}\|_1 + \|\boldsymbol{v}_{\mathrm{RBM}}\|_1\right) \leq \|\boldsymbol{v}_E\|_1 , \tag{5.25}$$

for every $\boldsymbol{v}_E = \hat{\boldsymbol{v}} + \boldsymbol{v}_{\mathrm{RBM}} \in \boldsymbol{V}_E$. Let us assume that (5.24) does not hold for any positive constant $c > 0$. Then, there exists a sequence $\{\boldsymbol{v}_{E,n}\}_{n\in\mathbb{N}} \subset \boldsymbol{V}_E$ such that

$$\|\boldsymbol{v}_{E,n}\|_1 = 1 \qquad \text{and} \qquad \|\varepsilon(\boldsymbol{v}_{E,n})\|_0 < \frac{1}{n} . \tag{5.26}$$

For each $\boldsymbol{v}_{E,n}$ we find $\hat{\boldsymbol{v}}_n \in \hat{\boldsymbol{H}}^1(\Omega)$ and $\boldsymbol{v}_{\mathrm{RBM},n} \in \boldsymbol{V}^{\mathrm{RBM}}$ with $\boldsymbol{v}_{E,n} = \hat{\boldsymbol{v}}_n + \boldsymbol{v}_{\mathrm{RBM},n}$. It holds that

$$\|\varepsilon(\hat{\boldsymbol{v}}_n)\|_0 = \|\varepsilon(\boldsymbol{v}_{E,n})\|_0 < \frac{1}{n} .$$

With Theorem 2.1.6 we conclude that $\hat{\boldsymbol{v}}_n \to \mathbf{0}$ in $[H^1(\Omega)]^2$. Since $\boldsymbol{V}^{\mathrm{RBM}}$ is three-dimensional, and since (5.25), implies the boundedness of $\boldsymbol{v}_{\mathrm{RBM},n}$, there exists a convergent subsequence $\{\boldsymbol{v}_{\mathrm{RBM},n_j}\}_{j\in\mathbb{N}}$. Hence, the subsequence $\{\boldsymbol{v}_{E,n_j} = \hat{\boldsymbol{v}}_{n_j} + \boldsymbol{v}_{\mathrm{RBM},n_j}\}_{j\in\mathbb{N}} \subset \boldsymbol{V}_E$ converges to some $\boldsymbol{v}_E \in \boldsymbol{V}^{\mathrm{RBM}}$, which implies that $\boldsymbol{v}_E = \mathbf{0}$ because $\boldsymbol{V}_E \cap \boldsymbol{V}^{\mathrm{RBM}} = \{\mathbf{0}\}$. This is a contradiction to the assumptions (5.26). □

Now, by using Korn's inequality, Poincarè's inequality and the inverse inequality we can show that

Lemma 5.5.2. *For all $\boldsymbol{v}_E \in \boldsymbol{V}_{II} = \boldsymbol{V}_E$ it holds that*

$$\|\boldsymbol{\varepsilon}(\boldsymbol{v}_E)\|_0^2 \approx \|\boldsymbol{\nabla}\boldsymbol{v}_E\|_0^2 \approx H^{-2}\|\boldsymbol{v}_E\|_0^2 . \tag{5.27}$$

Proof. The first equivalence follows from Korn's inequality on the space $\boldsymbol{V}_E$, Theorem 5.5.1, and from

$$\|\boldsymbol{\varepsilon}(\boldsymbol{v})\|_0 = \|\frac{1}{2}[\boldsymbol{\nabla}\boldsymbol{v} + (\boldsymbol{\nabla}\boldsymbol{v})^T]\|_0 \le \frac{1}{2}\|\boldsymbol{\nabla}\boldsymbol{v}\|_0 + \frac{1}{2}\|(\boldsymbol{\nabla}\boldsymbol{v})^T\|_0 = |\boldsymbol{v}|_1 \quad \forall \boldsymbol{v} \in [H^1(\Omega)]^2 .$$

The second equivalence is shown in two parts. First, we exploit Corollary 4.4.24 of [BS07]. It states that for any $v \in H^1(\Omega)$

$$\left(\sum_{T\in\mathcal{T}_H} \|v - I^h(v)\|_{0,T}^2\right)^{1/2} \le C_1 H |v|_{1,\Omega}$$

holds with I^h being a suitably element-wise defined interpolation operator and C_1 depending on ρ, being the constant of shape-regularity (see page (2.29)). Since we have that $\boldsymbol{V}_E \subset [H^1(\Omega)]^2$ and that $I_H(\boldsymbol{v}_E) = 0$, with the nodal linear interpolation I_H on $\mathcal{T}_H$, i.e., taking the nodal values at the vertices $v_i \in \mathcal{V}_H$, we arrive at

$$\left(\sum_{T\in\mathcal{T}_H} \|\boldsymbol{v}_E - I_H(\boldsymbol{v}_E)\|_{0,T}^2\right)^{1/2} = \left(\sum_{T\in\mathcal{T}_H} \|\boldsymbol{v}_E\|_{0,T}^2\right)^{1/2} = \|\boldsymbol{v}_E\|_{0,\Omega} \le C_1 H |\boldsymbol{v}_E|_{1,\Omega} .$$

In order to show the the second direction of the equivalence statement we make use of Theorem 4.5.11 in [BS07], i.e., an inverse estimate. We get for any $\boldsymbol{v}_h \in \boldsymbol{V}_h$

$$|\boldsymbol{v}_h|_1 \le \|\boldsymbol{v}_h\|_1 \le C_2 h^{-1}\|\boldsymbol{v}_h\|_0 = 2C_2 H^{-1}\|\boldsymbol{v}_h\|_0 , \tag{5.28}$$

with C_2 mainly depending on ρ. □

Next, note that any function $\boldsymbol{v}_E \in \boldsymbol{V}_E$ can be uniquely decomposed into $\boldsymbol{v}_E = \boldsymbol{v}_n + \boldsymbol{v}_\tau$ where $\boldsymbol{v}_n \in \boldsymbol{V}_n$ and $\boldsymbol{v}_\tau \in \boldsymbol{V}_\tau$.

Lemma 5.5.3. *Let $\boldsymbol{v}_E \in \boldsymbol{V}_E$ be given by $\boldsymbol{v}_E = \boldsymbol{v}_n + \boldsymbol{v}_\tau$ with $\boldsymbol{v}_n \in \boldsymbol{V}_n$ and $\boldsymbol{v}_\tau \in \boldsymbol{V}_\tau$. The following holds*

$$\|\boldsymbol{v}_E\|_0^2 = \|\boldsymbol{v}_n + \boldsymbol{v}_\tau\|_0^2 \approx \|\boldsymbol{v}_n\|_0^2 + \|\boldsymbol{v}_\tau\|_0^2 \tag{5.29}$$

with the constants $1-\bar{\gamma}$ and $1+\bar{\gamma}$, where $\bar{\gamma} = \frac{3}{7}$. The equivalence holds uniformly with respect to the mesh size h (without assuming shape-regularity of $\mathcal{T}_h$).

Proof. We estimate the CBS constant γ between the spaces $\boldsymbol{V}_n$ and $\boldsymbol{V}_\tau$ measured in the L_2-norm. By solving a generalized eigenvalue problem (see [KM09] for instance), we get for an arbitrary element $T \in \mathcal{T}_H$

$$|(\boldsymbol{v}_n\,,\, \boldsymbol{v}_\tau)_{0,T}| \le \gamma_T \|\boldsymbol{v}_n\|_{0,T} \|\boldsymbol{v}_\tau\|_{0,T} \tag{5.30}$$

with $\gamma_T \le \frac{3}{7}$. Further from [KM09, Lemma 2.5] we obtain

$$\gamma \le \max_{T\in\mathcal{T}_H} \gamma_T \le \frac{3}{7} = \bar{\gamma}\,. \tag{5.31}$$

Finally, the inequality on the right-hand side of (5.29) follows from

$$\begin{aligned}
\|\boldsymbol{v}_n + \boldsymbol{v}_\tau\|_0^2 &= \|\boldsymbol{v}_n\|_0^2 + \|\boldsymbol{v}_\tau\|_0^2 + 2(\boldsymbol{v}_n\,,\, \boldsymbol{v}_\tau)_0 \\
&\overset{(5.30),(5.31)}{\le} \|\boldsymbol{v}_n\|_0^2 + \|\boldsymbol{v}_\tau\|_0^2 + 2\gamma\|\boldsymbol{v}_n\|_0\|\boldsymbol{v}_\tau\|_0 \\
&\le (1+\gamma)(\|\boldsymbol{v}_n\|_0^2 + \|\boldsymbol{v}_\tau\|_0^2)\,,
\end{aligned}$$

where we have used that $2ab \le a^2 + b^2$ for $a,\, b \in \mathbb{R}$. Similarly the left-hand side of (5.5.3) can be shown. □

Now, let $I_{RT}^h : \boldsymbol{V}_{H,0}^{RT} \to \boldsymbol{V}_h$ be the interpolation operator from the Raviart-Thomas space $\boldsymbol{V}_{H,0}^{RT}$ to $\boldsymbol{V}_h$, defined by $I_{RT}^h(\boldsymbol{\varphi}_E^{RT}) = 2\varphi_E \boldsymbol{n}_E \in \boldsymbol{V}_n$. Note that $\Pi^{RT}(I_{RT}^h(\boldsymbol{v}_{RT})) = \boldsymbol{v}_{RT}$ for all $\boldsymbol{v}_{RT} \in \boldsymbol{V}_{H,0}^{RT}$ and additionally, note that $I_{RT}^h(\Pi^{RT}(\boldsymbol{v}_n)) = \boldsymbol{v}_n$ for all $\boldsymbol{v}_n \in \boldsymbol{V}_n$. That is, $\boldsymbol{V}_{H,0}^{RT}$ and $\boldsymbol{V}_n$ are isomorphic through I_{RT}^h and Π^{RT}.

Lemma 5.5.4. *For all $\boldsymbol{u}_h \in \boldsymbol{V}_h$ and all $\boldsymbol{u}_{RT} \in \boldsymbol{V}_{H,0}^{RT}$ we have*

$$\|\Pi^{RT}\boldsymbol{u}_h\|_0^2 \le c_{RT,1}\|\boldsymbol{u}_h\|_0^2\,, \tag{5.32}$$

$$\|I_{RT}^h\boldsymbol{u}_{RT}\|_0^2 \le c_{RT,2}\|\boldsymbol{u}_{RT}\|_0^2\,, \tag{5.33}$$

with $c_{RT,1} > 0$ *and* $c_{RT,2} > 0$ *being independent of* h.

Proof. First, we consider the l_2-norm of the operator Π^{RT} with respect to the space $\boldsymbol{V}_n$, represented through the matrix Π_h^{RT}. That is

$$\begin{aligned}
\|\Pi_h^{RT}\|_{l_2}^2 &:= \sup_{\|\boldsymbol{u}\|_{l_2}=1} \|\Pi_h^{RT}\boldsymbol{u}\|_{l_2}^2 = \sup_{\|\boldsymbol{u}\|_{l_2}=1} \sum_{E\in\mathcal{E}_H} \left[\left(\frac{\boldsymbol{u}_{E,1}+\boldsymbol{u}_{E,2}}{4} + \frac{\boldsymbol{u}_E}{2}\right)\cdot \boldsymbol{n}_E\right]^2 \\
&\overset{\text{C.S.}}{\leq} \sup_{\|\boldsymbol{u}\|_{l_2}=1} \sum_{E\in\mathcal{E}_H} \left\|\frac{\boldsymbol{u}_{E,1}+\boldsymbol{u}_{E,2}}{4} + \frac{\boldsymbol{u}_E}{2}\right\|_{l_2}^2 \\
&\overset{\Delta-\text{inequ.}}{\leq} \sup_{\|\boldsymbol{u}\|_{l_2}=1} \sum_{E\in\mathcal{E}_H} \frac{1}{2}\|\boldsymbol{u}_E\|_{l_2}^2 + \frac{1}{4}\sum_{v_i\in\mathcal{V}_H} m(i)\,\|\boldsymbol{v}_i\|_{l_2}^2 \leq \frac{1}{4}\max\{m,2\}\,.
\end{aligned}$$

Thereby, $m(i)$ denotes the vertex degree of vertex v_i and m is the maximum over all $m(i)$. Now, let M_h and M_{RT} denote the mass matrices with respect to $\boldsymbol{V}_n$ and $\boldsymbol{V}_{H,0}^{RT}$. Then, we obtain (5.32) through

$$\begin{aligned}
\|\boldsymbol{u}_h\|_0^2 &= \boldsymbol{u}^T M_h \boldsymbol{u} \geq \lambda_{\min}(M_h)\boldsymbol{u}^T\boldsymbol{u} \geq \frac{\lambda_{\min}(M_h)}{\|\Pi_h^{RT}\|_{l_2}^2}(\Pi_h^{RT}\boldsymbol{u})^T\Pi_h^{RT}\boldsymbol{u} \\
&\geq \frac{\lambda_{\min}(M_h)}{\lambda_{\max}(M_{RT})\|\Pi_h^{RT}\|_{l_2}^2}(\Pi_h^{RT}\boldsymbol{u})^T M_{RT}\Pi_h^{RT}\boldsymbol{u} \geq \frac{4\,\lambda_{\min}(M_h)}{m\,\lambda_{\max}(M_{RT})}\|\Pi^{RT}\boldsymbol{u}_h\|_0^2\,,
\end{aligned}$$

and since both, M_h and M_{RT}, are assembled from element matrices that have eigenvalues of the order $\mathcal{O}(H^2)$ on a quasi-uniform mesh. With $\boldsymbol{u} = (u_E)_{E\in\mathcal{E}_H}$, where $\boldsymbol{u}_{RT} = \sum_{E\in\mathcal{E}_H} u_E \boldsymbol{\varphi}_E^{RT}$, statement (5.33) follows from

$$\begin{aligned}
\|I_{RT}^h\boldsymbol{u}_{RT}\|_0^2 &= \|\sum_{E\in\mathcal{E}_H} u_E 2\varphi_E \boldsymbol{n}_E\|_0^2 = 4\boldsymbol{u}^T M_h\boldsymbol{u} \leq \frac{4\lambda_{\max}(M_h)}{\lambda_{\min}(M_{RT})}\boldsymbol{u}^T M_{RT}\boldsymbol{u} \\
&= \frac{4\lambda_{\max}(M_h)}{\lambda_{\min}(M_{RT})}\|\boldsymbol{u}_{RT}\|_0^2\,,
\end{aligned}$$

and from the same arguments as above. □

Now, let us define an auxiliary space $\bar{\boldsymbol{V}} := \boldsymbol{V}_\tau \times \boldsymbol{V}_{H,0}^{RT}$ with elements $\bar{\boldsymbol{v}} = (\boldsymbol{v}_\tau, \boldsymbol{v}_{RT})$. Further, we designate an auxiliary bilinear form $\bar{a}(.,.)$ on $\bar{\boldsymbol{V}}$, that is, for $\bar{\boldsymbol{u}}, \bar{\boldsymbol{v}} \in \bar{\boldsymbol{V}}$ we define

$$\begin{aligned}
\bar{a}(\bar{\boldsymbol{u}}, \bar{\boldsymbol{v}})) &:= 2\mu\Big\{H^{-2}(\boldsymbol{u}_\tau\,,\,\boldsymbol{v}_\tau)_0 \\
&\quad + H^{-2}(\boldsymbol{u}_{RT}\,,\,\boldsymbol{v}_{RT})_0 + \frac{\nu}{1-2\nu}(\operatorname{div}\boldsymbol{u}_{RT}\,,\,\operatorname{div}\boldsymbol{v}_{RT})_0\Big\}\,. \qquad (5.34)
\end{aligned}$$

Additionally, we set up the projection $\Pi : \bar{\boldsymbol{V}} \to \boldsymbol{V}_E$ by

$$\Pi(\bar{\boldsymbol{v}}) = \boldsymbol{v}_\tau + I_{RT}^h(\boldsymbol{v}_{RT}) . \tag{5.35}$$

With this, we are able to show the conditions of Theorem 3.4.1.

Theorem 5.5.5. *Let $\bar{\boldsymbol{V}}$ be defined by $\bar{\boldsymbol{V}} := \boldsymbol{V}_\tau \times \boldsymbol{V}_{H,0}^{RT}$ together with the auxiliary bilinear form $\bar{a}(.,.)$ given by* (5.34) *and the projection operator* (5.35). *In this setting, the preconditioner* (3.43) *satisfies* (3.44).

Proof. For B satisfying (3.44) we have to show the two conditions of the fictitious space lemma, Theorem 3.4.1. First, let $\boldsymbol{v}_E = \boldsymbol{v}_\tau + \boldsymbol{v}_n \in \boldsymbol{V}_E$ be arbitrary but fixed. Then, with $\bar{\boldsymbol{v}} = \left(\boldsymbol{v}_\tau, \Pi^{RT}(\boldsymbol{v}_E)\right) = \left(\boldsymbol{v}_\tau, \Pi^{RT}(\boldsymbol{v}_n)\right)$ it follows that $\Pi\bar{\boldsymbol{v}} = \boldsymbol{v}_E$ and

$$\|\bar{\boldsymbol{v}}\|_{\bar{A}}^2 = 2\mu \left\{ H^{-2}\|\boldsymbol{u}_\tau\|_0^2 + H^{-2}\|\Pi^{RT}(\boldsymbol{u}_n)\|_0^2 + \frac{\nu}{1-2\nu}\|\operatorname{div}\Pi^{RT}(\boldsymbol{u}_E)\|_0^2 \right\} .$$

Now, condition 1 of Theorem 3.4.1 follows from

$$\begin{aligned} H^{-2}\left(\|\boldsymbol{u}_\tau\|_0^2 + \|\Pi^{RT}(\boldsymbol{u}_n)\|_0^2\right) &\overset{(5.32)}{\lesssim} H^{-2}\left(\|\boldsymbol{u}_\tau\|_0^2 + \|\boldsymbol{u}_n\|_0^2\right) \\ &\overset{\text{Lemma 5.5.3}}{\lesssim} H^{-2}\|\boldsymbol{u}_E\|_0^2 \overset{\text{Lemma 5.5.2}}{\lesssim} \|\varepsilon(\boldsymbol{u}_E)\|_0^2 . \end{aligned}$$

Second, let $\bar{\boldsymbol{v}} \in \bar{\boldsymbol{V}}$ be arbitrary. Then

$$\|\Pi\bar{\boldsymbol{v}}\|_A^2 = 2\mu \left\{ \|\varepsilon(\Pi\bar{\boldsymbol{v}})\|_0^2 + \frac{\nu}{1-2\nu}\|\operatorname{div}\Pi^{RT}(\Pi\bar{\boldsymbol{u}})\|_0^2 \right\} .$$

Since $\Pi^{RT}(\Pi\bar{\boldsymbol{v}}) = \Pi^{RT}(\boldsymbol{v}_\tau + I_{RT}^h(\boldsymbol{v}_{RT})) = \boldsymbol{v}_{RT}$, condition 2 of Theorem 3.4.1 is satisfied through

$$\begin{aligned} \|\varepsilon(\Pi\bar{\boldsymbol{v}})\|_0^2 &\overset{\text{Lemma 5.5.2}}{\lesssim} H^{-2}\|\Pi\bar{\boldsymbol{v}}\|_0^2 = H^{-2}\|\boldsymbol{v}_\tau + I_{RT}^h(\boldsymbol{v}_{RT})\|_0^2 \\ &\overset{\text{Lemma 5.5.3}}{\lesssim} H^{-2}\left(\|\boldsymbol{v}_\tau\|_0^2 + \|I_{RT}^h(\boldsymbol{v}_{RT})\|_0^2\right) \overset{(5.33)}{\lesssim} H^{-2}\left(\|\boldsymbol{v}_\tau\|_0^2 + \|\boldsymbol{v}_{RT}\|_0^2\right) . \end{aligned}$$

Hence, the conditions of Theorem 3.4.1 are satisfied and therefore, the preconditioner B, given by (3.43), satisfies (3.44). □

From the previous theorem we see how to solve the (3.1) on the subspace $\boldsymbol{V}_{II}$. The inverse of $\bar{A}$ reduces to inverting a mass term, which can be done by preconditioning with its diagonal.

However, the inverse of $\bar{A}$ on $\boldsymbol{V}_{H,0}^{RT}$ is the problem of inverting

$$a_{RT}(\boldsymbol{u}_{RT}, \boldsymbol{v}_{RT}) := \mu \left\{ H^{-2}(\boldsymbol{u}_{RT}, \boldsymbol{v}_{RT})_0 + \frac{\nu}{1-2\nu}(\operatorname{div} \boldsymbol{u}_{RT}, \operatorname{div} \boldsymbol{v}_{RT})_0 \right\}, \tag{5.36}$$

An efficient solver for the latter problem can be designed by using the auxiliary space preconditioner of [HX07], or by using the robust algebraic multilevel iteration method developed in [KT11].

5.6 Numerical experiments

In order to confirm the convergence of the procedure corresponding to the error propagation (5.15) we perform the following numerical experiment. Consider the unit square $\Omega = (0,1)^2$ and the right-hand side $L(.) = 0$. We start the subspace correction algorithm with a random initial guess $\boldsymbol{u}_h^0$. In Table 5.1 we report the iteration count and the average convergence rate ρ that reduced the A-norm of the error by a factor of 10^8.

To come up with a reasonable choice for c_{III} in (5.15) we have computed $\lambda_{\max}(A_\varepsilon^{-1} A_{\text{div}})$ on small meshes and found it was bounded by 1.5. In order to satisfy condition (5.18), we choose $c_{III} = 2$ in all test cases. However, as we have discussed in Section 5.3 one usually needs some experimental data or an educated guess to choose c_{III} properly.

#DOF	98		338		1230		4802		18818	
	#it.	ρ	#it.	ρ	#it.	ρ	#it.	ρ	#it.	ρ
$\nu = 0$:	19	0.37	20	0.39	15	0.28	14	0.27	17	0.34
$\nu = 0.25$:	8	0.09	9	0.12	7	0.07	7	0.07	15	0.28
$\nu = 0.4$:	9	0.12	10	0.15	7	0.06	7	0.07	14	0.27
$\nu = 0.45$:	7	0.07	9	0.12	8	0.09	8	0.09	15	0.28
$\nu = 0.49$:	5	0.02	7	0.05	6	0.04	8	0.09	12	0.21
$\nu = 0.499$:	4	0.00	4	0.01	4	0.00	4	0.01	7	0.07
$\nu = 0.4999$:	3	0.00	3	0.00	3	0.00	3	0.00	4	0.01

Table 5.1: Iteration numbers (#it.) and convergence rates (ρ) of MSSC according to (5.15).

Additionally, we perform a numerical test to show that the preconditioner (5.22) is an efficient and robust preconditioner. We consider the problem with homogenous Dirichlet boundary conditions on the unit square $\Omega = (0,1)^2$. The number of PCG iterations for a residual reduction by a factor 10^8 are shown in Table 5.2. All subproblems are solved exactly. Additionally, we list the estimated condition numbers $\kappa(BA)$, obtained from the Lanczos process.

#DOF	242		1 058		4 418		18 050		72 962		293 378	
	#it.	κ	#it.	κ	#it.	κ	#it.	κ	#it.	κ	#it.	κ
$\nu = 0$:	1	1.00	1	1.00	1	1.00	1	1.00	1	1.00	1	1.00
$\nu = 0.25$:	8	1.41	8	1.48	8	1.53	9	1.55	9	1.57	9	1.57
$\nu = 0.4$:	10	1.90	11	2.19	12	2.38	12	2.49	13	2.57	13	2.62
$\nu = 0.45$:	11	2.11	12	2.61	14	3.01	15	3.25	15	3.41	15	3.52
$\nu = 0.49$:	10	1.90	11	2.54	14	3.31	16	3.97	17	4.39	17	4.69
$\nu = 0.499$:	9	1.98	10	1.98	11	2.13	14	2.99	15	3.83	17	4.51
$\nu = 0.4999$:	9	1.99	9	1.99	9	1.99	10	1.99	12	2.43	13	3.34
$\nu = 0.49999$:	9	1.99	9	1.99	9	2.00	9	2.00	9	2.00	10	2.00

Table 5.2: Iteration numbers (#it.) and condition numbers ($\kappa(BA)$) of the PCG-cycle with the preconditioner (5.22).

Chapter 6

Conclusion

6.1 Summary

In this book, the equations of linear elasticity are considered for which the focus lies on the pure displacement formulation. In order to complete the picture on possible variational formulations of the governing equations, also other possibilities to pose the problem in variational form are discussed. After discretization, the aim is to solve the arising linear system robustly with respect to mesh size and problem parameters such as incompressibility. The framework of subspace correction methods unites most of the state-of-the-art methods or concepts for solving discretizations of PDEs. Therefore, some known procedures are presented and their convergence properties are shortly addressed in terms of subspace correction methods.

In Section 3.6 innovative computations on the convergence of MSSC for two overlapping subspaces are derived. There, it is shown that the energy norm of the error propagation operator is given by the CBS constant of the operator with respect to the non-overlapping parts after an elimination of the overlap. The presented numerical results clearly underline the theoretical observations.

The AMG procedure based on computational molecules, which is presented in Chapter 4, is based on the concept of algebraic edges and vertices. Edge matrices determine the relation between the algebraic vertices and give rise to an approximation of the system matrix. In this chapter, a new measure for the strength of connectivity is introduced which reveals better insight on the nodal dependence. Additionally, a new way of computing the edge matrices is presented and the obtained approximation properties with respect to the element stiffness matrices are computed showing satisfying results. Further, the convergence of AMGm is investigated through an approximation of the condition number of the preconditioned operator. Thereby, it turns out that the CBS constant of the obtained space splitting is bounded under certain assumptions. If the smoother, which is in practical tests the Gauss-Seidel smoother, is compatible with the space splitting, the two-level procedure gives rise to a uniformly spectrally equivalent preconditioner. Remarks on an efficient parellelization of AMGm are made and finally numerical tests give evidence that the convergence of PCG using the W-cycle of AMGm is uniform. The experiments testify that AMGm outperforms its predecessor, published in [Kra08]. Further it outperforms BoomerAMG (see [HMY00, DSMYH06]), a state-of-the-art AMG solver, in terms of the convergence rate, which is due to a higher operator complexity of AMGm. At the end of this chapter, it is explained how to apply AMGm to stable DG discretizations of the linear elasticity equations in order to design an optimal solver, which is robust with respect to incompressibility. However, this is paid off by a higher computational complexity than standard MG for DG discretizations. The main results of this chapter have already been published in [KK10]

In Chapter 5 almost incompressible linear elastic materials are considered. Therefore, a locking-free discretization scheme based on reduced integration is utilized to obtain stable approximations of the solution. Based on an innovative space decomposition of the discrete space, an overlapping space decomposition using two subspaces is set up. Thereby, the subspaces are shown to be spanned by local basis functions leading to sparse representations of the system operator on the subspaces. An MSSC based on the subspace splitting is employed to define a preconditioner which shows optimal convergence properties in the case where the subproblems are solved exactly. However, since in practical applications the subspace problems have to be solved efficiently this issue is addressed as well. The first subproblem is independent of the problem parameters and therefore, it can be efficiently solved by means of any optimal order method for vector Laplace equations. The main problem is to devise a robust solution procedure for the second subproblem. In this work, an auxiliary space preconditioner is set up for this problem. The preconditioner is shown to be spectrally equivalent to the original operator by means of the fictitious space lemma. Finally, since for this preconditioner there do exist efficient and optimal solution procedures, one is able to implement efficient solvers for the subproblems which are of optimal order. The discussion on the preconditioner of this chapter has been recently published in the accepted article [KKZ11].

6.2 Outlook

Since there is always some work left over, we state some interesting open issues

- In Section 4.9 the application of AMGm to the DG formulation introduced in Subsection 2.4.3 is discussed. These considerations indicate that AMGm leads to a robust and optimal solution procedure. In order to find out about its practical properties and also possible limitations, the method has to be applied to the considered DG discretizations in numerical tests.

- In Section 5.4 the preconditioner (5.22) is introduced. The numerical results in Section 5.6 indicate an optimal convergence behavior of the PCG method. However, a theoretical confirmation of the robustness with respect to h and λ or ν, respectively, is an open issue. Based on the property of stable splittings, like (3.52), one could show convergence results. Therefore, one has to verify a condition similar to (3.52). If this can be achieved uniformly with respect to the considered parameters, a robust bound on the preconditioner would follow.

- In the numerical examples in Section 5.6 exact solutions of the subproblems are used. Nevertheless, numerical implementations of the preconditioners presented in 5.5 are of practical interest for the overall solution procedure derived in Chapter 5.

- Finally, note that the discussion in Chapter 5 is for plain strain elasticity problems, that is, for two-dimensional problems. The generalization of this for three-dimensional problems is an interesting issue.
  Note that, in [GR86] it is shown that a discretization of the mixed formulation of elasticity problems using piecewise linear functions plus face bubbles (cubic functions) for the velocities and piecewise constant pressure functions is stable for Stokes equations. Hence, this formulation can be used to set up a stable discretization scheme for linear elasticity problems posed purely in the displacements. There, the same space, as used for the velocities in the Stokes setting, is utilized for the FE discretization. Using again P_0 in the div-term yields an equivalent form to the mixed formulation, for which one might be able to show robust error estimates. Then, in order to devise a suitable overlapping space splitting, one has to find the equivalent subspaces to the ones presented in Chapter 5. Thereby, one can use that the commuting property $P_0 \operatorname{div} = \operatorname{div} \Pi^{RT}$ (see Lemma 5.1.1) holds also true for $d = 3$.

Bibliography

[ABCM01] D. N. Arnold, F. Brezzi, B. Cockburn, and L. D. Marini, *Unified analysis of discontinuous galerkin methods for elliptic problems*, SIAM J. Numer. Anal. **39** (2001), no. 5, 1749–1779.

[AC05] S. Adams and B. Cockburn, *A mixed finite element method for elasticity in three dimensions*, Journal of Scientific Computing **25** (2005), 515–521.

[AdDZ09] B. Ayuso de Dios and L. Zikatanov, *Uniformly convergent iterative methods for discontinuous Galerkin discretizations*, J. Sci. Comput. **40** (2009), no. 1-3, 4–36.

[AFW07] D. N. Arnold, R. S. Falk, and R. Winther, *Mixed finite element methods for linear elasticity with weakly imposed symmetry*, Math. Comput. **76** (2007), 1699–1723.

[AGKZ11] B. Ayuso, I. Georgiev, J. Kraus, and L. Zikatanov, *A subspace correction method for discontinuous galerkin discretizations of linear elasticity equations*, ESAIM: Math. Modelling and Numerical Analysis (M2AN) (2011).

[Arn81] D. N. Arnold, *Discretization by finite elements of a model parameter dependent problem*, Numer. Math. **37** (1981), 405–421.

[AV89] O. Axelsson and P. S. Vassilevski, *Algebraic multilevel preconditioning methods. I*, Numerische Mathematik **56** (1989), 157–177.

[AV90] ———, *Algebraic multilevel preconditioning methods, II*, SIAM J. Numer. Anal. **27** (1990), no. 6, 1569–1590.

[AW02] D. N. Arnold and R. Winther, *Mixed finite elements for elasticity*, Numer. Math **92** (2002), 401–419.

[Axe94] O. Axelsson, *Iterative solution methods*, Cambridge University Press, 1994.

[BCF+01] M. Brezina, A. J. Cleary, R. D. Falgout, V. E. Henson, J. E. Jones, T. A. Manteuffel, S. F. McCormick, and J. W. Ruge, *Algebraic multigrid based on element interpolation (AMGe)*, SIAM J. Sci. Comput. **22** (2001), no. 5, 1570–1592.

[BDY87] R. E. Bank, T. F. Dupont, and H. Yserentant, *The hierarchical basis multigrid method*, Numer. Math **52** (1987), 427–458.

[BF91] F. Brezzi and M. Fortin, *Mixed and hybrid finite element methods*, Springer series in computational mathematics, Springer-Verlag, 1991.

[BFM+04] M. Brezina, R. Falgout, S. MacLachlan, T. Manteuffel, S. McCormick, and J. Ruge, *Adaptive smoothed aggregation (αsa)*, SIAM Journal on Scientific Computing **25** (2004), no. 6, 1896–1920.

[BFM+06] M. Brezina, R. Falgout, S. MacLachlan, T. Manteuffel, S. McCormick, and J. Ruge, *Adaptive algebraic multigrid*, SIAM Journal on Scientific Computing **27** (2006), no. 4, 1261–1286.

[BMR82] A. Brandt, S. McCormick, and J. Ruge, *Algebraic multigrid (AMG) for automatic multigrid solution with application to geodetic computations*, Tech. report, Inst. Comp. Studies, Colorado State University, 1982.

[BPWX91] J. H. Bramble, J. E. Pasciak, J. Wang, and J. Xu, *Convergence estimates for multigrid algorithms without regularity assumptions*, Math. Comp. **57** (1991), no. 195, 23–45.

[Bra01] D. Braess, *Finite elements: Theory, fast solvers and applications in solid mechanics*, Cambridge University Press, 2001.

[Bre74] F. Brezzi, *On the existence, uniqueness, and approximation of saddle point problems arising from Lagrangian multipliers*, R.A.I.R.O., Anal. Numér. **2** (1974), 129–151.

[BS92a] I. Babuška and M. Suri, *Locking effects in the finite element approximation of elasticity problems*, Numer. Math. **62** (1992), 439–463.

[BS92b] I. Babuška and M Suri, *On locking and robustness in the finite element method*, SIAM J. on Numer. Anal. **29** (1992), no. 5, 1261–1293.

[BS07] S. C. Brenner and L. R. Scott, *The mathematical theory of finite element methods (texts in applied mathematics)*, 3rd ed., Springer, December 2007.

[CFH+98] A. J. Cleary, R. D. Falgout, V. E. Henson, J. E. Jones, T. A. Manteuffel, S. F. Mccormick, G. N. Miranda, and J. Ruge, *Robustness and scalability of algebraic multigrid*, SIAM J. Sci. Comput **21** (1998), 1886–1908.

[CFH+03] T. Chartier, R. D. Falgout, V. E. Henson, J. Jones, T. Manteuffel, S. McCormick, J. Ruge, and P. S. Vassilevski, *Spectral AMGe (ρAMGe)*, SIAM Journal on Scientific Computing **25** (2003), 1–26.

[Cho01] E. Chow, *An unstructured multigrid method based on geometric smoothness*, American Journal of Mathematics **65** (2001), 197–215.

[Cia78] P. G. Ciarlet, *Finite element method for elliptic problems*, North-Holland Publishing Company, Amsterdam, New York, Oxford, 1978.

[CV00] T. F. Chan and P. Vaněk, *Detection of strong coupling in algebraic multigrid solvers*, Multigrid Methods VI, Springer-Verlag, 2000, pp. 11–23.

[CXZ08] D. Cho, J. Xu, and L. Zikatanov, *New estimates for the rate of the method of subspace corrections*, Numerical Mathematics:Theory,Methods and Applications **1** (2008), 44–56.

[Deu01] F. Deutsch, *Best approximation in inner product spaces*, Springer, New York, 2001.

[DM04] R. G. Durán and M. A. Muschietti, *The Korn inequality for Jones domains*, Electronic Journal of Differential Equations **2004** (2004), no. 127, 1–10.

[DSMYH06] H. De Sterck, U. Meier Yang, and J. J. Heys, *Reducing complexity in parallel algebraic multigrid preconditioners*, SIAM Journal on Matrix Analysis and Applications **27** (2006), no. 4, 1019–1039.

[DW90] M. Dryja and O. B. Widlund, *Towards a unified theory of domain decomposition algorithms for elliptic problems*, Third International Symposium on Domain Decomposition Methods for Partial Differential Equations (Houston, TX, 1989), SIAM, Philadelphia, PA, 1990, pp. 3–21.

[ESGB82] M. S. Engelman, R. L. Sani, P. M. Gresho, and M. Bercovier, *Consistent vs. reduced integration penalty methods for incompressible media using several old and new elements*, Int. J. Numer. Meth. Fluids **2** (1982), no. 1, 25 – 42.

[Fal91] R. S. Falk, *Nonconforming finite element methods for the equations of linear elasticity*, Math. Comp. **57** (1991), no. 196, 529–550.

[FB97] J. Fish and V. Belsky, *Generalized aggregation multilevel solver*, Int. J. Numer. Meth. Engng. **40** (1997), 4341–4361.

[FV04] R. D. Falgout and P. S. Vassilevski, *On generalizing the algebraic multigrid framework*, SIAM Journal on Numerical Analysis **42** (2004), no. 4, 1669–1693.

[FVZ05] R. D. Falgout, P. S. Vassilevski, and L. T. Zikatanov, *On two-grid convergence estimates*, Numer. Linear Algebra Appl. **12** (2005), no. 5-6, 471–494.

[GK03] J. Gopalakrishnan and G. Kanschat, *A multilevel discontinuous galerkin method*, Numerische Mathematik **95** (2003), 527–550.

[GKM10] I. Georgiev, J.K. Kraus, and S. Margenov, *Multilevel preconditioning of Crouzeix-Raviart 3D pure displacement elasticity problems*, LSSC (I. Lirkov et al., eds.), LNCS, vol. 5910, Springer, 2010, pp. 103–110.

[GL81] A. George and J. W. H. Liu, *Computer solution of large sparse positive definite systems*, Englewood Cliffs, N.J.: Prentice-Hall, 1981.

[GO95] M. Griebel and P. Oswald, *On the abstract theory of additive and multiplicative schwarz algorithms*, Numer. Math. **70** (1995), no. 2, 163–180.

[GR86] V. Girault and P. A. Raviart, *Finite element methods for navier-stokes equations*, Springer-Verlag, Berlin Heidelberg New York Tokyo, 1986.

[Gri85] P. Grisvard, *Elliptic problems in nonsmooth domains*, Monographs and Studies in Mathematics, vol. 24, Pitman (Advanced Publishing Program), Boston, MA, 1985.

[Hac85] W. Hackbusch, *Multi-grid methods and applications*, Springer Berlin, 1985.

[HL03] P. Hansbo and M. G. Larson, *Discontinuous Galerkin and the Crouzeix-Raviart element: Application to elasticity*, Mathematical Modelling and Numerical Analysis **37** (2003), no. 1, 63–72.

[HLB79] T. Hughes, W. Liu, and A. Brooks, *Finite element analysis of incompressible viscous flows by the penalty function formulation*, Journal of Computational Physics **30** (1979), no. 1, 1–60.

[HLRS01] G. Haase, U. Langer, S. Reitzinger, and J. Schöberl, *Algebraic multigrid methods based on element proconditioning*, Int. J. comp. Math. **78** (2001), no. 4, 575–598.

[HMY00] V. E. Henson and U. Meier Yang, *BoomerAMG: a parallel algebraic multigrid solver and preconditioner*, Applied Numerical Mathematics **41** (2000), 155–177.

[HS52] M. R. Hestenes and E. Stiefel, *Methods of conjugate gradients for solving linear systems*, Journal of Research of the National Bureau of Standards **49** (1952), no. 6, 409–436.

[HT82] W. Hackbusch and U. Trottenberg (eds.), *Multigrid methods*, Lecture Notes in Mathematics, vol. 960, Springer, Berlin, 1982, Proceedings, Köln-Porz, 1981.

[HV01] V. E. Henson and P. S. Vassilevski, *Element-free AMGe: General algorithms for computing interpolation weights in AMG*, SIAM J. Sci. Comput. **23** (2001), 629–650.

[HW08] J.S. Hesthaven and T. Warburton, *Nodal discontinuous galerkin methods: Algorithms, analysis and applications*, Texts in Applied Mathematics, vol. 54, Springer, 2008.

[HX07] R. Hiptmair and J. Xu, *Nodal auxiliary space preconditioning in H(curl) and H(div) spaces*, SIAM J. Numer. Anal. **45** (2007), no. 6, 2483–2509.

[Joh87] C. Johnson, *Numerical solution of partial differential equations by the finite element method*, Cambridge University Press, Cambridge, 1987.

[Jon81] P. W. Jones, *Quasiconformal mappings and extendability of functions in sobolev spaces*, Acta Mathematica **147** (1981), no. 1, 71–88.

[JV01] J. E. Jones and P. S. Vassilevski, *AMGe based on element agglomeration*, SIAM J. Sci. Comput. **23** (2001), 109–133.

[Kik86] N. Kikuchi, *Finite element method in mechanics*, Cambridge University Press, 1986.

[KK10] E. Karer and J. K. Kraus, *Algebraic multigrid for finite element elasticity equations: Determination of nodal dependence via edge-matrices and two-level convergence*, Int. J. Numer. Meth. Engng. **83** (2010), no. 5, 642–670.

[KKZ11] E. Karer, J. K. Kraus, and L. Zikatanov, *A subspace correction method for nearly singular linear elasticity problems*, 20th Internat. Conf. on Domain Decomposition Methods, UC San Diego, Springer-Verlag, 2011, accepted, pp. 1–8.

[KM87] M. Kočvara and J. Mandel, *A multigrid method for three-dimensional elasticity and algebraic convergence estimates*, Applied Mathematics and Computation **23** (1987), no. 2, 121–135.

[KM09] J. K. Kraus and S. D. Margenov, *Robust algebraic multilevel methods and algorithms*, Radon Series Comp. Appl. Math., vol. 5, Walter de Gruyter, 2009.

[KO89] V. A. Kondratiev and O. A. Oleinik, *On Korn's inequalities*, C. R. Acad. Sci. Paris Sér. I Math. **308** (1989), no. 16, 483–487.

[Kra08] J. K. Kraus, *Algebraic multigrid based on computational molecules, 2: Linear elasticity problems*, SIAM J. Sci. Comput. **30** (2008), no. 1, 505–524.

[KS06] J. K. Kraus and J. Schicho, *Algebraic multigrid based on computational molecules, 1: Scalar elliptic problems*, Computing **77** (2006), 57–75.

[KT11] J. K. Kraus and S. K. Tomar, *Algebraic multilevel iteration method for lowest order Raviart-Thomas space and applications*, International Journal for Numerical Methods in Engineering **86** (2011), no. 10, 1175–1196.

[KV06] T.V. Kolev and P. S. Vassilevski, *AMG by element agglomeration and constrained energy minimization interpolation*, Numerical Linear Algebra with Applications **13** (2006), no. 9, 771–788.

[KvRD+99] J. Kabel, B. van Rietbergen, M. Dalstra, A. Odgaard, and R. Huiskes, *The role of an effective isotropic tissue modulus in the elastic properties of cancellous bone*, Journal of Biomechanics **32** (1999), 673–680.

[KW88] S. Kayalar and H. Weinert, *Error bounds for the method of alternating projections*, Mathematics of Control, Signals, and Systems (MCSS) **1** (1988), 43–59.

[LM07] R. D. Lazarov and S. D. Margenov, *CBS constants for multilevel splitting of graph-laplacian and application to preconditioning of discontinuous galerkin systems*, Journal of Complexity **42** (2007), no. 4-6, 498–515.

[LRS00] U. Langer, S. Reitzinger, and J. Schicho, *Symbolic methods for the element preconditioning technique*, Proc. SNSC Hagenberg, U. Langer and F. Winkler, eds., Springer, 2000.

[LWC09] Y.-J. Lee, J. Wu, and J. Chen, *Robust multigrid method for the planar linear elasticity problems*, Numerische Mathematik **113** (2009), 473–496.

[LWXZ07] Y.-J. Lee, J. Wu, J. Xu, and L. T. Zikatanov, *Robust subspace correction methods for nearly singular systems*, Math. Models Methods Appl. Sci. **17** (2007), no. 11, 1937–1963.

[MH94] J. E. Marsden and T. J. R. Hughes, *Mathematical foundations of elasticity*, Dover Publications Inc., New York, 1994, Corrected reprint of the 1983 original.

[MP96] J. F. Maitre and O. Pourquier, *Condition number and diagonal preconditioning: comparison of the p-version and the spectral element methods*, Numer. Math. **74** (1996), no. 1, 69–84.

[MW10] K. A. Mardal and R. Winther, *On the construction of preconditioners for systems of partial differental equations*, Efficient preconditioning methods for elliptic partial differential equations (Owe Axelsson and Janos Karatson, eds.), Bentham Science Publishers, 2010, http://simula.no/research/sc/publications/Simula.sc.862.

[MW11] ———, *Preconditioning discretizations of systems of partial differential equations*, Numer. Linear Algebra Appl. **18** (2011), no. 1, 1–40.

[Néd80] J.-C. Nédélec, *Mixed finite elements in* $\mathbb{R}^3$, Numer. Math. **35** (1980), no. 3, 315–341.

[Néd86] ———, *A new family of mixed finite elements in* $\mathbb{R}^3$, Numer. Math. **50** (1986), no. 1, 57–81.

[Nep91] S. V. Nepomnyaschikh, *Mesh theorems on traces, normalizations of function traces and their inversion*, Russian Journal of Numerical Analysis and Mathematical Modelling **6** (1991), no. 3, 223 – 242.

[Nep92] ———, *Decomposition and fictitious domains methods for elliptic boundary value problems*, Fifth International Symposium on Domain Decomposition Methods for Partial Differential Equations (Norfolk, VA, 1991), SIAM, Philadelphia, PA, 1992, pp. 62–72.

[Nep07] ———, *Domain decomposition methods*, Lectures on Advanced Computational Methods in Mechanics (J. Kraus and U. Langer, eds.), Radon Series on Computational and Applied Mathematics, vol. 1, de Gruyter, Berlin, 2007, pp. 89–159.

[Not05] Y. Notay, *Algebraic multigrid and algebraic multilevel methods: a theoretical comparison*, Numerical Linear Algebra with Applications **12** (2005), no. 5-6, 419–451.

[Pec08] C. Pechstein, *Finite and boundary element tearing and interconnecting methods for multiscale elliptic partial differential equations*, Ph.D. thesis, Johannes Kepler University Linz, Austria, 2008.

[Pec12] ______, *Finite and boundary element tearing and interconnecting solvers for multiscale problems*, Springer, 2012, to appear.

[RS80] M. Reed and B. Simon, *Functional analysis, revised and enlarged edition*, Methods of Modern Mathematical Physics, vol. 1, Academic Press, 1980.

[RS87] J. W. Ruge and K. Stüben, *Algebraic multigrid (AMG)*, Multigrid Methods (Philadelphia) (S. F. McCormick, ed.), Frontiers Appl. Math., vol. 3, SIAM, 1987, pp. 73–130.

[Sch97] J. Schöberl, *NETGEN - An advancing front 2d/3d-mesh generator based on abstract rules*, Comput. Visual. Sci. **1** (1997), no. 1, 41–52.

[Sch99a] J. Schöberl, *Multigrid methods for a parameter dependent problem in primal variables*, Numer. Math. **84** (1999), no. 1, 97–119.

[Sch99b] J. Schöberl, *Robust multigrid methods for parameter dependent problems*, Ph.D. thesis, Institute for Computational Mathematics, J.K. University Linz, 1999.

[Sin09] A. Sinwel, *A new family of mixed finite elements for elasticity*, Ph.D. thesis, Johannes Kepler University Linz, Austria, 2009.

[SS07] J. Schöberl and A. Sinwel, *Tangential-displacement and normal-normal-stress continuous mixed finite elements for elasticity*, `http://www.ricam.oeaw.ac.at/publications/reports/07/rep07-10.pdf`, 2007, RICAM-report 2007-10, Johann Radon Institute for Computational and Applied Mathematics (RICAM).

[Stü01] K. Stüben, *An introduction to algebraic multigrid*, in et al. [TOS01].

[SV85] L. R. Scott and M. Vogelius, *Norm estimates for a maximal right inverse of the divergence operator in spaces of piecewise polynomials*, RAIRO Modél. Math. Anal. Numér. **19** (1985), no. 1, 111–143.

[TOS01] U. Trottenberg, C. Oosterlee, and A. Schüller, *Multigrid*, Elsevier Academic Press, London, 2001.

[TW05] A. Toselli and O. Widlund, *Domain decomposition methods - algorithms and theory*, Springer-Verlag, Berlin Heidelberg New York, 2005.

[Vas08] P. S. Vassilevski, *Multilevel block factorization preconditioners: Matrix-based analysis and algorithms for solving finite element equations*, Springer, 2008.

[VBM01] P. Vaněk, M. Brezina, and J. Mandel, *Convergence of algebraic multigrid based on smoothed aggregation*, Numer. Math. **88** (2001), 559–579.

[VMB96] P. Vaněk, J. Mandel, and M. Brezina, *Algebraic multigrid based on smoothed aggregation for second and fourth order problems*, Computing **56** (1996), 179–196.

[WFTS04] H. Waisman, J. Fish, R. S. Tuminaro, and J. N. Shadid, *The generalized global basis (GGB) method*, Int. J. Numer. Meth. Engng. **61** (2004), 1243–1269.

[WFTS05] ———, *Acceleration of the generalized global basis (GGB) method for nonlinear problems*, J. Comput. Phys. **210** (2005), no. 1, 274–291.

[Xu92] J. Xu, *Iterative methods by space decomposition and subspace correction*, SIAM Review **34** (1992), no. 4, 581–613.

[Xu96] ———, *The auxiliary space method and optimal multigrid preconditioning techniques for unstructured grids*, Computing **56** (1996), 215–235.

[XZ02] J. Xu and L. Zikatanov, *The method of alternating projections and the method of subspace corrections in hilbert space*, Journal of the American Mathematical Society **15** (2002), no. 3, 573–597.

[YKvR+99] G. Yang, J. Kabel, B. van Rietbergen, A. Odgaard, R. Huiskes, and S. C. Cowin, *The anisotropic Hooke's law for cancellous bone and wood*, Journal of Elasticity **53** (1999), no. 2, 125–146.

[Yse86] H. Yserentant, *On the multi-level splitting of finite element spaces*, Numer. Math **49** (1986), 379–412.

[Zik08] L. T. Zikatanov, *Two-sided bounds on the convergence rate of two-level methods*, Numer. Linear Algebra Appl. **15** (2008), no. 5, 439–454.

[ZSTB10] Y. Zhu, E. Sifakis, J. Teran, and A. Brandt, *An efficient parallelizable multigrid framework for the simulation of elastic solids*, ACM Transactions on Graphics **29** (2010), no. 2, 1–18.

MIX
Papier aus verantwortungsvollen Quellen
Paper from responsible sources
FSC® C105338

Printed by Books on Demand GmbH, Norderstedt / Germany